T0353580

Deep Learning in Visual Computing
Explanations and Examples

Hassan Ugail

Director, Centre for Visual Computing
University of Bradford
Bradford, UK

CRC Press
Taylor & Francis Group
Boca Raton   London   New York

CRC Press is an imprint of the
Taylor & Francis Group, an **informa** business

A SCIENCE PUBLISHERS BOOK

First edition published 2022
by CRC Press
6000 Broken Sound Parkway NW, Suite 300, Boca Raton, FL 33487-2742

and by CRC Press
4 Park Square, Milton Park, Abingdon, Oxon OX14 4RN

© 2022 Taylor & Francis Group, LLC

CRC Press is an imprint of Taylor & Francis Group, an Informa business

No claim to original U.S. Government works

Library of Congress Cataloging-in-Publication Data (applied for)

ISBN: 978-0-367-54962-6 (hbk)
ISBN: 978-0-367-54963-3 (pbk)
ISBN: 978-1-003-09135-6 (ebk)

DOI: 10.1201/9781003091356

Typeset in Times New Roman
by Radiant Productions

Dedication

To all those deep learning algorithms helping humanity

Preface

Deep learning is an artificially intelligent entity that teaches itself to make predictions following a training phase through an intensive data driven algorithm. Deep learning, through the adoption of artificial neural networks, functions much like the human brain to classify and analyse data. Deep learning has enabled to solve many challenging problems in the area of visual computing. From object detection to image classification for diagnostics, deep learning has shown the power of artificial deep neural networks in solving real world visual computing problems with super-human accuracy. Thus, with most traditional image processing and computer vision techniques effectively being replaced by deep learning techniques it has proved to have the potential to solve many of the most challenging problems in visual computing.

This book provides an insight into the world of deep learning applicable to the field of visual computing. The book provides easy to understand explanations on the theory and architectures behind deep learning. To help the reader better grasp the concepts of deep learning and how it can be applied in real life, the book is then supplemented with a number of real-life examples which range from face recognition to cancer image classification. The book is written to serve as a recipe for thought provoking and as a source of inspiration for enthusiasts in developing and implementing their own artificial intelligence powered projects in the field of visual computing.

Acknowledgments

This book is a result of many years of research into the topic of machine learning by the author, his colleagues, post-doctoral researchers and research students at the Centre for Visual Computing at the School of Engineering and Informatics at University of Bradford, UK. Both fundamental and applied research into a topic of the nature described in this book requires both human capital and computational resources. Much of the funding for such resources came from research grants and sponsorships. The list of funders who helped me generously to pursue this work with my team members is exhaustive and undoubtedly memorable. Moreover, it takes a fair amount of dedication and effort to put together a book of this nature and usually work on a book of this type falls outside the regular lines of the day to day academic activities. As a result, a support base outside the immediate academic environment is essential to complete a book of this nature. I feel blessed to have received such support from my dear family members, friends, colleagues and collaborators. Without their support and continued encouragement, this book would not have seen the light of the day. My heartfelt thanks, therefore, goes to all of them. Thank you.

Contents

CHAPTER 1

Introduction

In September 2017, Apple released its iPhone X. With it came the Face ID—Apple's advanced facial recognition-based biometric system—embedded within the hardware and the software of the phone. Shortly, I managed to lay my hands on an iPhone X. To use facial recognition as an identification system—and to gain access to various Apps, including the App that lets me into my bank account—I first had to enroll my face. The enrolling process involved following a simple set of instructions where I had to present my face at certain poses and angles to the phone's camera. Within a few minutes of enrolling my face, I was able to start enjoying the convenience of simply looking at my phone to unlock it and use it.

Onto another story. A few years back, I was on a regular walk at the local park with my son, who was 4-year-old then. Children with their small size, large eyes, chubby cheeks, and a display of innocence and exuberance on their faces, appear to naturally attract adults' attention and affection. As we were strolling along a small pathway in the park, we encountered a middle-aged lady who briefly stopped and said "Hello" to my son, while giving him a social smile. My son gently returned the greeting. Since then, we often see her in the local park, and we exchange regular greetings. My son probably will never forget her face. The most common visual object we humans process is the face. The human brain has an amazing ability to process faces, to recognise them—often decades following the first glimpse of a given face.

There is one key parallel one could draw between how the human brain distinguishes faces and how an algorithmically based face recognition system such as Apple's Face ID works. For a human, recognising a face is a trivial task. Though all the faces are similar—with two eyes, a nose, and a mouth in a uniform configuration—we can still identify and recognise faces in many different angles and lighting conditions, and we do this often with ease. It is believed that humans perform image recognition through a template matching process in which the perceived objects—for example, faces—are stored in a long term memory in the form of a discrete set of features that truly represent a given object. And, when a new object is presented, its features can be compared with those existing in the long term memory to retrieve the best match. Thus, two distinct processes can be identified here, the first being the learning process in which the brain teaches itself about specific objects through features. For example, in the story of my son at the park, his brain learnt the face of the lady he met in the park. Then, in the second process, the input face is compared to the stored face templates to find an exact match.

Interestingly, Apple's Face ID identifies and discriminates between faces in much the same way as we humans identify and recognise faces in our brain. At the enrolling stage, Face ID reads the face from various positions and angles and ultimately converts the physical features of the face into a mathematical representation—often this would be a few hundred floating-point numbers per face. For recognition, the infrared sensor on the phone reads the face, and computes the mathematical representation corresponding to the face and makes a "distance-wise" comparison with the stored template face(s). Again, one can see, like the brain, the face matching system first learns about the face, and then it is compared to the template faces to see if there is a match verifying a given threshold value of accuracy.

The fundamental idea behind object recognition–both in the computer world and the human brain—is information compression or the automatic formation of useful representations from data. Foundationally, we can refer to this process of representational learning as deep learning. A good example of this is the way we often represent the structure of an atom, as shown in Figure 1.1. Whilst an atom contains electrons that surround a nucleus composed of protons and neutrons, its structure is never as formative, as shown in Figure 1.1. In fact, the position of an electron at any given time is never predictable, and only a probability estimate of where it is located around the nucleus is what we can obtain. However, presenting the structure of an atom, as shown in Figure 1.1 helps us to form a piece of representational learning about atoms, molecules, and macroscopic matter.

The brain of a human contains well over 100 billion neurons. Together these neurons build a colossal network that is parallel and distributed. Functional activities such as seeing and interpreting images are carried using these networks. Similarly, in the digital algorithmic world, artificial neural architectures can be created to mimic how the brain learns and performs complex tasks. In a visual computing setting, as an example, one can imagine providing millions and millions of images to a machine learning algorithm as training data. The algorithm figures out the unique patterns in the image data and ultimately puts them

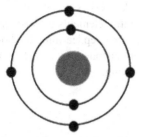

Figure 1.1: A structural representation of the carbon atom.

into specific categories through representational learning. The important point to make a note of here is that the neural network algorithm need not be explicitly coded to detect certain features visible or classifiable by a human. Rather, the algorithm is left to its own devices to figure out the complex patterns and classify the data appropriately. Thus, with sufficient data, the neural network learns about the objects on its own through a training and testing process.

At a very high level, deep neural networks are composed of encoders and decoders. The function of the encoder is to find useful patterns in the data and come up with a representational form of learning. Similarly, the function of the decoder is to generate high-resolution data from the representations where the generated data is new examples or expressive knowledge. And, fundamentally, this is how deep learning works.

With the recent explosion of powerful computational tools and sensor data, deep learning has become a tool that is central to machine learning. Deep learning has found its use in great arrays and varieties. Virtual assistants, such as Siri and Alexa, biometric face recognition, chatbots, image analysis tools, and computer-aided diagnostic systems, are being actively driven by deep learning-assisted models and algorithms. In fact, the entire field of artificial intelligence has recently taken centre stage due to the explosive developments in deep learning algorithms and the availability of pre-trained deep learning models. Furthermore, much of the problems in the domain of visual computing are being addressed using deep learning methods and tools. From basic image analysis tasks to complex disease diagnosis exercises, deep learning has taken a central role in almost all visual computing- related problems. Therefore, a book solely on the accumulated knowledge and recent developments in deep learning in the domain of visual computing deserves to be authored.

This book outlines the methods and techniques of deep learning applied to problems arising from the domain of visual computing. The first section of this book is devoted to explaining deep learning—starting from the fundamental concepts to the mechanisms of crafting deep learning models. The book covers sufficient theory on deep learning for the user to grasp the essential elements in deep learning. To supplement the theory and knowledge on deep learning, the book also covers several practical examples to demonstrate how it can be successfully utilised in the realm of visual computing.

CHAPTER 2

The Foundations of Deep Learning

This chapter provides a brief introductory material on deep learning. We provide the fundamental concepts and techniques one must bear in mind for understanding deep learning from the point of view of attempting to solve problems in visual computing.

Introduction

Artificial intelligence (AI) is defined to be a technique that enables a machine or an algorithm to mimic human behaviour. Much of the engine of artificial intelligence is driven by methods and techniques of Machine Learning (ML) with deep learning being part of it. Deep learning allows computational models to learn and represent data in a manner mimicking how the human brain perceives and understands multi-modal information. Though deep learning has become very popular only recently, its history is rather long.

Deep learning is a mechanism by which a machine algorithm can learn by example. It is a mechanism to obtain an optimal configuration to a model so that the desired output can be obtained from a set of input data. Mathematically, one can think of this process as obtaining the desired function $g(y_1, y_2,..., y_n)$ that maps to an input function $f(x_1, x_2,..., x_m)$. There exist many ways one can obtain such relationships—starting from linear approximations to making use of complex non-linear forms. Often, explicit mathematical relationships between variables associated with physical laws or phenomena can be formulated. For example, if we take the well-known Newton's second law of motion which states that the force F applied to an object of mass m is the product of m and the acceleration, which is the rate at which the velocity of the object changes. This is simply formulated as a linear relationship such that $F = ma$. Therefore, given the input values of m and a, one could easily predict the force that must be exerted. Similarly, if we take Einstein's famous equation, $E = mc^2$, we can see that a non-linear relationship exists between the energy of a piece of matter m and the speed of light c.

For a selected set of physical phenomena, and for more straightforward real-life situations such correlations between variables can be inferred which can then be explicitly written down in a linear or a non-linear form. However, much of the real-world problems are too complicated or too dynamic to be able to find such explicit relationships between the associated variables.

Consider the relationship between the input and output of a system as shown graphically in Figure 2.1. The input and output relationship, shown in Figure 2.1 (a), is clearly linear which can be modelled using a function of the form $F(x) = ax + b$, where a and b are feature values (or parameters) in the model space. In some sense, the function $F(x)$ with specific values a and b being able to closely approximate the dataset in Figure 2.1(a) is a learned model. If the input and output pattern of the model now changes, as shown in Figure 2.1(b), the linear model no longer is valid. However, we can resort to a model of the form $F(x) = ax^3 + bx^2 + cx + d$ or $F(x) = Ce^{Ax}$ (where a, b, c, d and C and A are the parameters of the corresponding non-linear model) which would now be more suitable for modelling the data shown in Figure 2.1(b).

Though the linear and non-linear models are suitable for modelling simple datasets, for real-life applications, this sort of functional modelling is challenging to obtain. For example, for a visual computing task like identifying an apple in an image, seeking a linear combination of pixels that can map the image of an apple is unattainable. Even with a simple combination of non-linear functions, such a task can be daunting where there could be several huge and often conflicting parameters to deal with. As a result, one must look for a generic methodology through which complex and multidimensional datasets that apply to real-

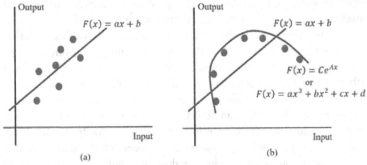

Figure 2.1: Illustration of the structure of a perceptron and how it functions.

life applications can be modelled. This is one area whereby machine learning, especially the technique of deep learning, can help us. Such models can take sufficiently large data as input and process them to extract the general patterns in the data using high-level information or knowledge. They are geared to mimic how humans infer objects and how humans extract knowledge from them. As a result, a plausible avenue for teaching a machine about the world is to mimic the human brain— particularly how the neurons utilise data for learning and inferring.

The Perceptron

The first machine learning computational model, inspired by the neurons in the brain, was proposed by Warren McColloch and Walter Pitts, both of whom worked in the field of computational neuroscience back in the 1940s. They wanted to understand how the human brain could produce intricate patterns by using interconnected neurons. They tried to use it as a foundation for their proposed perceptron model, initially proposed by Rosenblatt in 1958, who was working at the Cornell Aeronautical Laboratory at the time.

The McColloch and Pitts perceptron is a mathematical model based on the working concept of a biological neuron. Neurons are interconnected nerve cells within the brain and are collectively known as neural networks. They function by processing and transmitting information via electrical and chemical signals. A biological neuron is stimulated when an action potential is generated because of a change in the ion concentration across the cell membrane. Generally, there is a higher concentration of sodium ions in the extracellular space while there is a higher concentration of potassium ions within the intracellular space. During an action potential, ions are transported back and forth across the neuronal membranes, causing an electrical change that transmits the nerve impulse.

Similar to the functioning of a human neuron, the perceptron neural model receives a series of incoming signals x_1, x_2, x_3, which would either be excitatory or inhibitory. The signals can be weighted through w_1, w_2, w_3, so that their effect can be adjusted as necessary. Finally, the weighted sum can be computed. If the weighted sum of the incoming signals is at a chosen threshold, the model gives an output of 1, if not the output is 0. Thus, the model bases its output decision on the input signals, i.e., x_1, x_2, x_3, by performing a weighted sum, which in turn generates a binary output, i.e., 0 or 1, as shown in Figure 2.2.

The perceptron is a classification technique and can be widely utilised in visual computing applications, for example, in computer

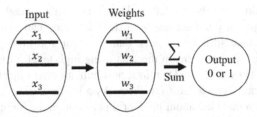

Figure 2.2: Illustration of the structure of a perceptron and how it functions.

vision and image recognition. Suppose we have a picture and we want the algorithm to tell us if the picture is that of an apple or not, you can use classification for that. The input could be the pixels of the image, the weights of the perceptron can be derived by using a dataset of apples, and the output will be a binary classification telling us if the input image is an apple or not. However, just like the function, a single neuron is very limited, there are many real-world problems that cannot be solved using binary classifiers. Thus, the perception idea can be extended to artificial neural networks (ANN), where a network of neurons can be connected together.

Artificial Neural Networks

Essentially an ANN is a collection of individual perceptrons forming a network of neurons connected to achieve parallel signal processing across various parts of the network. Between the perceptron units, they contain weights to control the effect of one another. Figure 2.3 shows the basic structural arrangement of an ANN. A key component of an ANN is the hidden layers that are placed between the input and output of the network. The hidden layers help in assigning non-linear weights to the

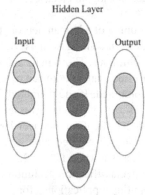

Figure 2.3: Representation of a multilayer neural network with a hidden layer.

input and direct them to the output, i.e., they help to apply non-linear mathematical functions to specific parts of the network to produce a desired final output. These functions are often referred to as activation functions. For example, in the case of a face detection task, one hidden layer can be tuned to identify the eyes in the input image, another can be tuned to identify the nose, and so on. Together, the network can then recognise the existence of a face in the input image.

Specifically, in an ANN-based on the multilayer perceptron concept, each neuron is connected to the neurons in the previous and the following layer. The connections that exist between the input and the output layer are defined by a number of hidden layers which are connected by activation functions. The activation function A usually takes the form,

$$A = F(\sum_{i=1}^{n} w_i x_i + b), \tag{2.1}$$

where, w_i is the weight assigned to the neuron i and b is a bias value with n being the number of neurons. The input data defines the number of neurons in the input layer. The hidden layers, with the help of the activation function, perform multiple feature extraction for classification. Finally, all the hidden layers provide a feature set as an output. The number of neurons of the output is equivalent to the number of classes to which the input data can be divided into. The accuracy of the output, i.e., the probability of an input belonging to a given class, is measured using a loss function, otherwise known as a Softmax.

The activation function itself can take specific forms, though largely this should not matter too much. Some common forms of the activation function utilised in an ANN are, the linear function, where $F(x) = ax$, the rectified linear unit (ReLU), where $F(x) = \max(0, x)$, the hyperbolic tan, where $F(x) = tan(x)$, the logistic function, where $F(x) = \dfrac{1}{1 + e^x}$ and the inverse of tan, where $F(x) = \dfrac{1}{1 + tan(x)}$.

Similarly, in the network, for each neuron, a bias is added, which is usually a binary value. Bias units are singular in the sense that they are not connected to the previous layer. The bias function enables the activation function to fire in a given direction, e.g., right, left, up, or down. The introduction of bias in the model adds flexibility to it. It ensures the model gets trained accordingly.

Once a neural network is crafted, consisting of initial weights, the error function tells us how far the model is from the desired predictions required of it. To efficiently obtain the optimal values for the weights of the network, one can perform the backpropagation. There are several stages in the backpropagation algorithm. First, with the initial arrangement of the weights, the model can generate its initial prediction. The error function can then be computed. The backpropagation algorithm then computes the rate of change of error with respect to individual weights so that the error minimisation can be undertaken iteratively. These iterations or optimisation runs, which backtracks the weights and updates them in the backpropagation algorithm, are often referred to as epochs.

Thus, the neural network takes input data and passes the inputs through the network to the output. The network must evaluate how well it performs through the loss function S such that, for example, the mean square error,

$$S = 1/n \sum_{i=1}^{n}(Y_i^1 - Y_i^2)^2, \qquad (2.2)$$

is minimised. Here, Y_i^1 represents the expected output and Y_i^2 represents the predicted outcome of the network. Through a series of iteration using backpropagation, the network adjusts its weights and biases to ensure the loss function is minimised. Hence, from an input, through a series of adjustments of the weights and biases the network produces an expected output. In short, an ANN is an optimised system modelling the relationship between given inputs and the expected outputs.

Designing an ANN for solving a given task is not often a straightforward process. It requires careful thought and experimentation. The choice of the input, the number of hidden layers, the choice of activation functions, the nature of the Softmax, the optimisation algorithm chosen, and the number of epochs for training the network all require due consideration. This process is referred to as hyperparameter tuning.

The multilayer perception models are excellent in providing good approximations for predictive modelling. It is worth noting that the very nature of these models means that the various layers are fully connected where each perceptron is connected to every other perceptron. As a result, for the visual computing-related problems, where the requirements are mostly related to images in the pixel domain, the number of parameters can grow exponentially very quickly. In addition to this, these types of models do not encode spatial information efficiently, which often makes them prohibitive for use to solve common visual computing problems. And to overcome such issues, researchers have come up with more sophisticated models. Thus, the popular Convolutional Neural Network (CNN) models are the current favourite in the area of visual computing.

Convolutional Neural Networks

A CNN is a three-dimensional network structure in which the neurons of one layer are not always connected to those in the next layer. The neurons in each layer are tasked with a specific function for analysing smaller sections of data. Thus, a CNN first performs a convolution which involves scanning a small section of the input data to create a specific map. The second step involves pooling which attempts to reduce the dimensionality of the features. In the case of an image, the convolution operation scans the image and analyses small parts of the image to produce a feature set. The pooling operation then reduces this feature set to a manageable number while retaining crucial information.

Thus, a CNN can be explained as a combination of layers, where each layer is responsible for a different task, such as convolution or pooling. Such networks work by allowing an image to pass through as input, then go through a set of layers containing convolutions, pooling, and other fully connected layers, to finally provide an output of a single class of a set of possible classes for the image, as shown in Figure 2.4.

At first, when CNNs were introduced, there were similarities to normal neural networks, where CNNs would mainly contain neurons that hold learnable weights and biases. Each neuron normally receives several inputs, it then performs a dot product, and finally has the option to follow it with a non-linearity. Like the normal artificial neural networks, the CNNs also have a loss function, such as Softmax, in the last layer. Moreover, methods that are used for learning regular neural networks still apply in CNNs.

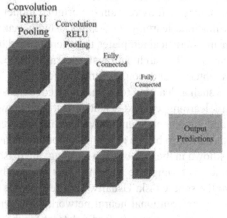

Figure 2.4: Illustration of the generic structure of a convolutional neural network.

The recent developments in CNNs improve the overall capabilities with the passing of every year, with more complex and computationally demanding network models being proposed. Recently, the top leading model for the ImageNet challenge was the Inception-v4 model architecture, which was able to achieve 3.08% top error on the ImageNet challenge through using 75 trainable layers, overcoming the ResNet and GoogLeNet that were the previous champions in image classification. However, one of the downsides of the latest state-of-the-art is the increasing training run time.

Thus, a layer in CNN would have neurons arranged in 3 dimensions—width, height, and depth. A CNN normally contains different layer types stacked in different orders and quantities to produce the best architecture. These layers include a convolutional layer that would contain several filters to extract features from the image data. This pooling layer would perform down-sampling on the data to focus on the most relevant patterns. This fully connected layer would be attached to the end of the network in order to give the network the ability to classify data samples. Finally, at the end of most network structures, it can be found that a Softmax classifier is added to provide the output class.

Transfer Learning

In deep learning, often model definition and model training is an extensive task. Often deep learning is very data-hungry. As a result, for an initial model training, a large dataset is essential—however, one way to work around this is to utilise a technique called transfer learning. Transfer learning enables a model trained for one task to be adapted for another, without having to fully re-train the original model.

Traditional machine learning algorithms usually make predictions on future data using statistical templates that were trained on previously collected training data. Research on transfer learning has been attracting more and more interest and consideration since 1995, where many names were used such as learning to learn, knowledge transfer, inductive transfer, multitask learning, context-sensitive learning, knowledge-based inductive bias, and incremental/cumulative learning. At present, transfer learning methods appear to be rife in visual computing-related literature.

As already known in the area of deep learning, the reason for the use of transfer learning is to improve learning. This is done by leveraging knowledge from the source task. Usually, very few people tend to build and train an entire convolutional neural network from scratch, mainly because it is not common or easy to find a dataset for the required data type with sufficient size. At the same time, it is more common to train a CNN on an extensive dataset, such as the Image-net, and then use the CNN either as an initialisation or a fixed feature extractor for the task of interest.

Classification Schemes

CNNs typically consist of a distributed network with many layers. These are responsible for feature extraction. The final layer, which is a Softmax, does the actual classification. This gives CNNs the unique ability to train feature extraction instead of leaving feature definition to predetermination. These extracted features from different layers of CNN have also been used to train separate classifiers, such as the Support Vector Machines (SVM), k-nearest neighbours (kNN), and random forest (RF) for predictions.

The Support Vector Machine (SVM) is a type of classification model. This model aims to find the hyperplanes with the largest "margin" separating data into classes, i.e., the distance between the closest points in opposite classes to the hyperplane is maximised. If data is separable, the problem of placing a hyperplane in the data space is reduced. Often data may not be separable, and so a penalty for having a data point on the "wrong" side of the hyperplane must be defined mathematically. Let C define the hyperparameter, which represents how much weight will be permissible to the misclassifying points. Then, a large C heavily penalises misclassified points while encouraging hyperplanes that fit the training data more closely. On the other hand, a smaller C allows for more misclassified points in the hope that this will result in a more generalisable classifier. As such, parameter C is the primary tool to overcome overfitting in SVM and is usually determined through the use of validation. SVM has three types of kernel functions K which transform the input data into another Euclidean space. The kernels could be linear, non-linear and it may be Gaussian (with the use of Radial Base Functions).

The k-Nearest Neighbour (kNN) is a relatively simple classification model which uses a known dataset to classify new data. The new data point is classified based on the class with the majority representation among the k nearest neighbours. The idea behind this is straightforward. For example, points of the same class should have similar characteristics. As a result, the features of points of the same class should be sufficiently similar to one another. Therefore, such points should be "near" to other points of the same class. The notion of "nearness" here refers to the locations of points in the feature space as well as the distance between them. Distance often is measured using the Euclidean representation. When the value of the k term is high, the classification is more resistant to be influenced by the outlier data points. Hence k represents a hyperparameter that can be chosen for effective validation. These classifiers are simple and intuitive. However, they tend to struggle with scaling because classification requires revisiting all the data points for every new data point to be considered. Additionally, the classifier must have some prior knowledge of all the training data points to classify the new data points. Naturally, this results in a large amount of memory for storing all of the training data along with the classification protocols.

The Random Forest (RF) is a classification method that creates a set of decision trees. The trees are randomly selected subsets of the main training set. It then aggregates the votes from different decision trees to decide the final class of the test object. RF is a popular ensemble method. It uses the technique of bagging which relates to the generation of a new training dataset by selection and random selection of a test set from the highest-ranked tests at an internal node of a tree. It requires little parameter tuning, and can be used to rank the importance of features naturally. RF runs efficiently on large datasets and reports accurate classification performance. These characteristics make RF a popular classification algorithm in machine learning.

Summary

As opposed to explicit programming for physical models, deep learning helps to generate computational systems to extract patterns and knowledge from data. This has a close resemblance to how the human brain recognises and infer patterns for complex decision making. Deep learning provides a computational framework for obtaining a desired output based on a given set of input data.

A perceptron is the simple form of a machine learning unit. It can undertake linear approximations of complex functions very efficiently. Similarly, an artificial neural network is based on a collection of perceptrons with input, output, and hidden layers all connected with weights, activation functions, and biases. An artificial neural network can help incorporate non-linear mathematical function units into the system so that input data is accurately approximated using a smaller set of features. With its feature-based output, it can help with image analysis tasks and has proven to be useful in visual computing applications. However, a drawback with the traditional multi-perceptron-based artificial neural networks is that for complex tasks such as face recognition, the number of feature parameters can get very large, and the network becomes computationally inefficient.

Convolutional Neural Networks are at present the de-facto machine learning technique for visual computing applications. This is proven to be useful in many areas, and it has shown its strength in solving challenging tasks such as winning many ImageNet competitions. The network is sparsely connected, rather than fully connected all the way through, with relatively smaller number weights. With the convolution and pooling functions and the loose connectivity of the network, it has advantages over traditional multi-perceptron-based networks. As a result, it is excellent for pattern recognition and matching and therefore, has been proven to be excellent for solving many challenging visual computing related problems.

Further Reading

LeCun, Y., B. Y. Bengio, and G. Hinton. Deep learning. Nature 521.7553, pp. 436–444, 2015.

Nielsen, M. Neural Networks and Deep Learning, http://neuralnetworksanddeeplearning.com/2013.

CHAPTER 3

Deep Learning Models for Visual Computing

Deep learning seeks to emulate and learn complex arrangements in datasets using multiple learning layers of processing units (neurons) to recognise them as the human brain does. Deep learning algorithms learn very complex data via a series of repeated or multiple artificial neurons by manipulating the parameters associated with the neuron to produce the desired output. As described in the previous chapter, deep learning is a neural network with multiple hidden layers. This can be associated with how humans think to solve a complex problem by applying intelligence to it. Similarly, deep learning does that by using deep layers to figure out an efficient solution to a given classification problem.

Deep learning is successfully recognised in diverse areas of applications such as image and video analysis, object tracking, human action recognition, image stylisation, and face recognition. Its popularity these days has to do with increasing computing power such as the emergence of graphic processor units, the lower cost of hardware, and the increase in the speed of network connectivity. Additionally, the popularity of deep learning has to do with the abundance of data available to train and test the algorithms. The prime success of deep learning applicable to visual computing problems is the emergence of efficient convolutional neural networks. Based upon that, an abundance of pre-trained models has been made available recently.

Convolutional Neural Networks

Recently, convolutional neural networks (CNNs) have had a significant impact in the field of visual computing. This is due to their ability to learn complex features using nonlinear multi-layered architectures. Although originating in the early 1990s, CNNs were forsaken by the research community due to the assumption that feature extraction using gradient descent will always overfit as a result of the local minima in the objective function of the related optimisation problem. However, its remarkable success in the ImageNet competition in 2012 altered this pessimistic view associated with it. Today, state-of-the-art deep models are used in almost all visual computing applications including, but not limited to, detection, recognition, classification, and information retrieval.

Supervised learning is one of the most common forms of machine learning. To appreciate the concept, consider the problem of building a system that can classify banknotes as real or counterfeit. One will collect a large dataset of images (real and fake notes), then at training time, the machine is shown a photograph and its category, the machine then produces an output in the form of two scores, one for each category. Ideally, the output from the machine is assumed to be good if it gives the best score to the target category. However, we are almost sure the machine cannot produce such output before training. Fortunately, one can track the performance of the machine at the time of training, by computing a cost function that measures the difference (error) between the machine's output and that of the target scores. To reduce the error, the machine then perturbs its internal adjustable parameters. Normally these parameters, called weights, define a mapping of the input to output. This training procedure of showing an image, estimating its scores, comparing it to the target, and adjusting the weights goes on iteratively until a zero error, or very minimal acceptable error is achieved. Once training has been accomplished, the weights of the machine now have intelligent values which can be used to map a new (unseen) image of the banknote to a category with minimal error. The above scenario explains the basic working of a feed-forward neural network.

Interestingly, research has shown that CNNs efficiently learn generic visual features. Thus, these features can be used directly with simple classifiers to solve visual computing problems. This approach, known as off-the-shelf-feature extraction, has been used by several researchers to achieve promising results in visual computing tasks. Researchers advise that, rather than training CNNs from scratch, transfer learning should be the first approach to solving a visual computing task. Likewise, some studies suggest that, for a dataset with a small number of images, the off-the-shelf feature extraction technique outperforms training a network from scratch.

As discussed in the previous chapter, a neural network like the one shown in Figure 3.1, can perform the stated cognitive task of classifying banknotes, only if trained to do so. In the example of Figure 3.1, the network has 2 inputs, 2 hidden layers with 4 neurons and 2 outputs, so it has ($2^2 \times 2^2 \times 2^2$) = 12 connections. Consequently, each connection has

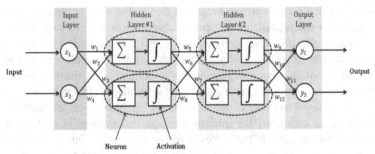

Figure 3.1: Feed-forward artificial neural network.

a weight. As described in the example above, training involves iterative tweaking of the weights based on the output error. It is also obvious that every neuron from layer l is connected to the output of every neuron from layer $l-1$. This is of course the key attribute of a fully connected neural network (NN), and it is called a feed-forward NN since the output of a layer becomes the input of the next layer. Passing an input and getting the predicted output is called the forward pass. This involves computing the total net input to each hidden layer neuron, passing the result through a nonlinear (activation) function, and then repeating the process with the next layer of neurons.

Various activation functions have been used in CNNs. Today, the rectified linear unit (ReLU) is the most popular activation function, which is simply a ramp function $f(z) = max(z,0)$. During the backward pass, weights are tuned to minimise the error. This is achieved by a technique known as backpropagation. The procedure computes the partial derivative of the error concerning the weights, achieved by working backward. This computation then indicates by what amount the error decreases or reduces as a result of a small change in the weights. Subsequently, the weights are adjusted in the opposite direction of the computed gradient. After adjustment of the weights, the output error changes, thus before the next iteration, the partial derivatives have to be recomputed once again. Due to the large number of parameters involved in a neural network, the algorithm often overfits—a phenomenon in which the model performs excellently on training dataset—classification with minimum error. However, it fails to generalise on a new unseen dataset. Methods for avoiding overfitting include the use of a large training dataset, stopping the training as soon as performance on a validation set starts to get worse, regularisation, and dropout.

Just like the feed-forward neural network, a CNN can be defined as a function q composed of a sequence of simpler functions p such that,

$$q = p_l \circ p_{l-1} \circ \ldots \circ p_1, \tag{3.1}$$

where each function p defines a mapping of input x_{l-1} of the previous layer to its output x_l expressed as,

$$x_l = f(\Sigma w_l x_{l-1} + b),\qquad(3.2)$$

where f is an activation function, w_l are weights and b is the bias.

Thus, a Convolutional Neural Network (CNN) is a type of artificial neural network that takes into consideration the spatial structure of the input data. To ensure shift and distortion invariance, CNNs combine three architectural ideas—shared weights, local receptive fields, and spatial or temporal subsampling. Weight sharing refers to the procedure of applying repetitive (shared) tiles of neurons across space. This results in lesser parameters to optimise, and consequently, increases the learning efficiency. As it is impractical to join the neurons to all neurons in a previous layer, especially in large dimensional data (for instance images), the local receptive field ensures the connection of neurons to only local regions in the input volume. Subsampling and local averaging enhance the efficiency of the algorithm by decreasing the resolution of the feature map, and therefore, decreasing the sensitivity of the output to shifts and distortions. In general, CNNs take the shape of a 3D structure tensor, having W × H × D dimensions where W (width) and H (height) are spatial dimensions, whereas D is the feature dimension. Specifically, the structure of CNNs makes them most appropriate for image, speech, and time-series tasks. More specifically, the algorithm earned its name due to the convolution operation that is used to apply the weights to the input.

CNNs were first discovered in the 1990s. However, unfortunately, despite their breakthrough in document recognition, they were forsaken by researchers due to the assumption that neural networks will always overfit or get trapped in local minima. However, research has shown that the existence of a local minimum within the search space of the optimisation problem is not necessarily that much of a problem. The difference in performance is only slightly affected when the local minima are minimally non-optimal. Moreover, a conventional method of avoiding local minima is by perturbing the stability of the algorithm, so that it suddenly hops out of the local optimum. Varying the learning rate by gradually and repeatedly increasing and decreasing it, reduces the stability of the algorithm, thus giving the algorithm the ability to jump out of a local optimum. Additionally, the stochastic gradient descent (SGD) optimisation algorithm also introduces some sort of noise or randomness, which helps the algorithm to jump from local minima. In essence, the obvious difference between CNNs of today and those of the 90s include the availability of extremely large datasets such as the ImageNet, faster computation realised by parallel processing ability of GPUs, advanced techniques for initialising weights at the start of training, and the use of simple and easy ways to differentiate activation functions.

The Architecture of a CNN

The structure of a typical CNN is composed of multiple layers that fall into three broad categories. They are the convolution layers (CONV), subsampling layers (POOL), and fully connected layers (FC). Furthermore, the activation function can also be considered as a layer in the architecture. Usually, a combination of the aforementioned layers is arranged in a specific manner with the sole goal of transforming the input of the network into a useful representation that gives an output.

This is the fundamental building block of CNN. As stated earlier, it derives its name from the mathematical (convolution) operator. This layer computes the scalar product between the weights of neurons and a small region of the input volume. The neurons are arranged as a stack of 2-dimensional filters/kernels that extend the depth of the input volume. Hence they are 3D structured. During the forward pass, each kernel is convolved across the width and height of the input volume to produce a 2D feature map, as shown in Figure 3.2. Thus, these feature maps are the outputs of the convolution operation at each spatial operation. In comparison to the feed-forward NN, here filters represent neurons that activate when they come across visual features such as edges. As

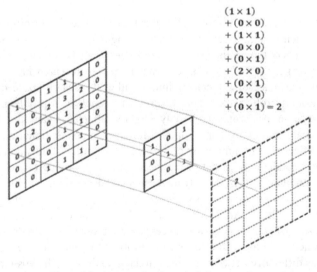

Figure 3.2: An example of a convolution operation.

discussed earlier, CNNs use local connectivity to reduce complexity. Hence each neuron is connected to a local region whose spatial dimension is defined by the filter size known as the receptive field of the neuron, and its depth is always equal to the depth of the input volume.

Hence, for a $256 \times 256 \times 3$ input image, if the receptive field is 3×3, then each neuron in the CONV layer will have a total of $3 \times 3 \times 3 = 27$ connections, and 1 bias parameter. Obviously, the connectivity is spatially local but full along with the input depth. Subsequently, the size of the feature map (i.e., the output) is computed using three hyperparameters—depth, zero padding, and stride. Depth refers to the number of filters deployed. The more the number of filters, the more the information retrieved since each filter learns to look for a specific feature. Stride defines a pattern used to slide the filter across the input, $S = 1$ means the filter can move one pixel at a time along with the input. Zero padding defines the number of zero pixels placed around the input volume to keep the spatial size of the output volumes. One can compute the spatial size of the output using,

$$O = \left\{ \frac{I - F + 2P}{S} \right\} + 1, \qquad (3.3)$$

where I is the spatial size of the input, F is the filter size, P is the number of zero paddings and S is the stride. Hence, if applied to input images of size $[224 \times 224 \times 3]$ and assuming the neurons are having a receptive field of 3×3 in size, depth $K = 64$, a single stride $S = 1$, and zero paddings $P = 1$, then one gets $\dfrac{224 - 3 + 2}{1} + 1 = 224$. This means that the output volume of this particular CONV layer will have a size $[224 \times 224 \times 64]$. Consequently, there will be $224 \times 224 \times 64 = 3211264$ neurons, each having $3 \times 3 \times 3 = 27$ weights and 1 bias. Interestingly, rather than having 3211264×27 weights and 3211264 biases, the concept of weight sharing makes all the neurons on one slice share the same weight and bias. Hence, the number of weights and biases drastically reduces to 1728 and 64, respectively.

As mentioned earlier, in neural networks, the activation function plays a significant role in introducing nonlinearity to the output of a neuron. Introducing this nonlinearity makes the neural network a universal function approximator, thereby, giving it the ability to understand various types of relationships. The most effective and commonly used activation function for CNNs is the rectified linear unit (RELU). This involves the element-wise application of a zero thresholding function, $f(x) = \max(o, x)$ where x is the input of the neuron. Compared to other activation functions, CNNs with ReLUs train several times faster. The activation layer does not introduce additional parameters to its input. Furthermore, it does not change the dimension of the input. In the architecture, activation layers are placed after every CONV layer. Additionally, networks with more than one FC layer also deploy it except for the last fully connected layer.

Pooling layers are usually inserted between successive CONV layers. Their primary function is to consistently reduce the number of parameters and consequently decrease the computational complexity of the network by reducing the spatial size of the feature maps. Hence, they summarise the output of the neighbouring neurons. For every 2D slice of the feature map, the most common type of pooling operation called MAX-POOLING usually takes the maximum of each 2 × 2 region, thus discarding 75% of the activations as shown in Figure 3.3.

Thus, the pooling operation does not introduce new parameters. Rather, it leads to shrinkage of the first and second dimensions of the feature map. The operation takes two parameters, the stride S and the spatial dimension F. Hence the pooling operation reduces a feature map from $W_1 \times H_1 \times D$ to $W_2 \times H_2 \times D$ dimension. Here W_2 and H_2 are computed using,

$$W_2 = \frac{W_1 - F}{S} + 1, \; H_2 = \frac{H_1 - F}{S} + 1. \tag{3.4}$$

Interestingly, this operation introduces translational invariance with respect to elastic distortions.

The fully connected layer has neurons that have a full connection to the previous layer's activation, and unlike the CONV and POOL layers, the FC has a 2D dimension. They are typically configured to output the network's predicted label/classes. Hence the FC is usually the last layer of the network. In the work that won the 2012 ImageNet Large Scale Visual Recognition Competition (ILSVRC), 3 FC layers were used, and since then this has been a rule of thumb among most researchers. Intuitively, flattening the 3D feature maps at the end of the computation gives us an avenue for interpreting the learned spatial invariant features.

The most popular arrangement used by researchers starts with the image input layer, and ends with an FC (decision) layer, in between

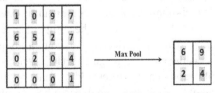

Figure 3.3: Max-pool with a 2 × 2 filter having a stride of 2.

these two are repeated stacks of CONV-RELU layers followed by POOL layers, then a few FC-RELU layers. This layered pattern can be described mathematically as,

$$INPUT \Rightarrow N\{M(CONV \Rightarrow RELU) \Rightarrow POOL\}$$

$$\Rightarrow K(FC \Rightarrow RELU) \Rightarrow FC.$$

Usually, the number of CONV-RELU layers that appear before POOL is within the range $0 < N < 4$, and the combinations of variables M and K are usually greater than 1. A typical example is the popular VGG16 model.

Methods of Training CNNs

Generally, there are three ways of deploying CNNs—training a network from scratch, fine-tuning an existing model, or using off-the-shelf CNN features. The latter two approaches are referred to as transfer learning. Since training CNNs from scratch, using the backpropagation algorithm involves the automatic learning of millions of parameters, this approach requires an enormous amount of data, often in millions. More so, this data-hungry nature of CNNs consequently demands large computational power. Furthermore, the procedure involves the adjustment of several hyperparameters. Thus, people rarely train an entire network from scratch.

Fine-tuning involves transferring the weights of the first layers learned from a base network to a target network, and then continuing the backpropagation using the new dataset. Hence, the target network is trained using the new dataset for a specific task, usually different from that of the base network. Fine-tuning is recommended when the new dataset is moderately large (tens to hundreds of thousands) and very different from the base network's dataset. Using the weights of the old network to initialise helps the backpropagation algorithm, and so leading to relatively fast automatic learning of more specific features.

In situations where the dataset is rather small, say few hundreds, even fine-tuning the weights may result in overfitting. However, since CNNs efficiently learn generic image features, it is then possible to directly use a trained network as a fixed feature extractor. Hence, features from new data are extracted by projecting them onto activations of a specific layer of the pre-trained network. After that, the learned representations are fed into simple classifiers to solve the task at hand. This approach, known as off-the-shelf feature extraction, has been used by several researchers to achieve promising results.

Data augmentation is the simplest way to combat the overfitting problem encountered by deep neural networks. This technique works by artificially enlarging the dataset size through various techniques such as changing the image orientation by flipping (which produces the mirrored image) and rotating the original images, which subsequently result in a new image. This ensures the learning algorithm learns features from lots of data with different orientations.

Deep CNN Models

The deep CNNs were trained on a subset of the ImageNet database, which is used in the ImageNet Large-Scale Visual Recognition Challenge (ILSVRC). These models were trained on more than a million images and can classify images into 1000 object categories. There are two types of models. They are, a series model of a network with layers arranged one after the other such as AlexNet and VGG architectures. A DAG model has layers arranged as a directed acyclic graph. They have a more complex architecture in which layers have inputs from multiple layers and outputs to multiple layers. Examples include GoogleNet, Inception, ResNet, and IncResNet architectures. We now consider the following architectures as our models for the fine-tuning strategy.

The VGG

VGG was proposed by the Visual Geometry Group at the University of Oxford. VGG16 refers to the VGG model with 16 weight layers, while VGG19 refers to the VGG model with 19 weight layers. It has a similar architecture with AlexNet with more convolutional layers. VGG16 has 13 convolutional layers, whereas VGG19 has 16 convolutional layers and then both are followed by rectification and pooling layers, and three fully connected layers. All convolutional layers use small 3×3 filters, and the network performs only 2×2 pooling

The VGG architecture has a receptive field of size 224×224. The output layer is a Softmax which performs the prediction on 1000 classes. From the input layer to the last max-pooling layer is the feature extraction part of the model. The rest of the network contains the classification part of the model.

The VGG is a publicly available model that was trained using 2.6 million face images of 2622 unique subjects. The model is configured to take a fixed-sized [$224 \times 224 \times 3$] RGB image as an input, as a form of pre-processing. All the images used are centre-normalised. The network is made of a stack of 13 convolutional layers with filters having a uniform receptive field of size 3×3 and a fixed convolution stride of 1 pixel. As

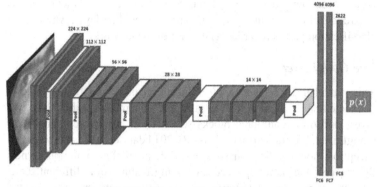

Figure 3.4: Architecture of the VGG-Face model.

shown in Figure 3.4, groups of these convolution layers are followed by five max-pooling layers. Finally, the convolutional layers are then followed by three fully connected layers, FC6, FC7, and FC8. The first two have 4096 channels, while FC8 has 2622 channels which are used to classify the 2622 identities. In addition to centre normalisation, the model's implementation also incorporates 2D alignment.

Given an input image X_0 represented as a tensor $X_0 \in \mathbb{R}^{H \times W \times D}$ where H is the image height, W is the width and D the colour channels, and a pre-trained L layered ConvNet expressed as a series of functions $q_L = p_1 \rightarrow p_2 \rightarrow \cdots p_L$. To fully investigate and evaluate which layer yields the best age descriptor, the activation of five layers—the last two convolution layers (conv5_2, conv5_3), the last max-pool layer (pool5), and the first two fully connected layers FC6 and FC7 of the VGG-Face model are used as separate feature channels. The choice of layers has been restricted to the top 5 layers, because going further down yields extremely huge dimensions that will result in no significant gains even after reducing the dimension.

Due to the large number of dimensions of the extracted features, ranging from 4096 in FC7 to 100352 in conv5_2, there is a need to reduce the feature size, thus removing redundant information. Moreover, it is a well-known fact that, for n observations and p features, the regression estimate is not well-defined in a situation where $p > n$.

AlexNet

AlexNet is the winner of the ImageNet 2012 challenge that popularised CNNs. It contains five convolutional and pooling layers and three fully connected layers, including local response normalisation (ReLU) layers and dropouts. It operates on $227 \times 227 \times 3$ input images which are cropped randomly from 256 × 256 images. The feature extraction part is regarded from the input layer to the last max-pooling layer, whereas the classification part is regarded as the rest of the network.

The GoogLeNet

This model has 22 layers with 9 inception units and finally, a fully connected layer before the output. The inception module has two layers and 6 convolutional blocks which is an intrinsic component of GoogLeNet. It is trained on the ILSVRC2014 dataset. It uses the so-called Inception modules that consist of multiple parallel convolution kernels that process the same input, concatenating features at a different scale. Also,the features extracted by the pooling operations are concatenated by the Inception module. Using the inception module, GoogLeNet can achieve high accuracy with limited computational cost.

The Inception-v3

The Inception is larger, deeper, and slower than GoogLeNet, but more accurate on the original ILSVRC data set. It is a 48-layer deep model and consists of multiple convolutional and pooling layers in which the outputs are concatenated. It operates on 229 × 229 × 3 input images. It migrated from fully connected to sparsely connected architectures. It had more nonlinearity capability by including the 1 × 1 factorised convolutional neural networks followed by the rectified linear unit (ReLU). Also, a 3 × 3 convolutional layer was employed. Auxiliary logits with a combination of an average pool, convolutional 1 × 1, fully connected, and Softmax activation is applied to preserve the low-level detail features and tackles the vanishing gradient problem in the last layers.

The ResNet

The residual network architecture was developed by Microsoft. The connections of ResNet enable the training of deeper networks. ResNet-50 is 50 layers deep, and ResNet-101 is 101 layers deep. It operates on 224 × 224 × 3 input images. The residual networks seek to increase the network's depth without problems affecting the results. The central idea of residual networks is based on the introduction of an identity function between layers. In conventional networks, there is a nonlinear function $y = H(x)$ between layers (underlying mapping), as shown on the left of Figure 3.5. In residual networks, we have a new nonlinear function $y = F(x) + id(x) = F(x) + x$. Here $F(x)$ is the residual, as shown in Figure 3.5.

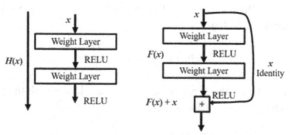

Figure 3.5: A typical CNN and shortcut connections of ResNet architecture.

This modification (called shortcut connections) allows important information to be carried from the previous layer to the next layers. Doing this avoids the problem of the vanishing gradient (many neurons become useless during the training of such deep networks). The advantages of the residual networks are that they manage to increase the depth of the network without increasing the number of parameters to optimise, thus accelerating the training speed of very deep networks. They also reduce the effects of the disappearance of the gradient problem, thus improving the accuracies obtained.

The Inception-ResNet

The Inception-ResNetcombines a high-efficiency inception module of GoogLeNet with the residual connections of ResNets. Inception-ResNet-v2 is 164 layers deep. It operates on $229 \times 229 \times 3$ input images. It is a variation of the Inception V3 model that borrows some ideas from the articles on the Microsoft ResNets networks, used in the previous section. Residual connections include shortcuts in models that, as mentioned, allow researchers to train even deeper networks that achieve better performance. This has also allowed a significant simplification of the Inception blocks.

Standard Performance Measurements

The performance of neural networks can be evaluated according to standard performance terms such as true positive (TP), false positive (FP), true negative (TN), false negative (FN), accuracy (ACC), precision (P), and sensitivity (S). The standard performance measurements are formulated as depicted in the following equations.

Accuracy evaluates the overall effectiveness of the classifier such that,

$$Accuracy = \frac{(TP + TN)}{(TP + FN + FP + TN)}. \qquad (3.5)$$

Sensitivity evaluates the effectiveness of the classifier to identify positive labels such that,

$$Sensitivity\ (Recall) = \frac{TP}{TP + FN}.$$ (3.6)

Precision evaluates the class agreement of data labels with the positive labels given by the classifier such that,

$$Precision = \frac{TP}{TP + FN}.$$ (3.7)

These four classifiers constitute a confusion matrix shown in Figure 3.6 for the case of binary classification.

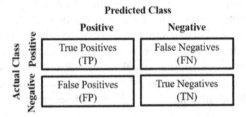

Figure 3.6: A Confusion Matrix for the Binary Classification.

Cross-validation ensures the data used during the training do not participate during the testing phase. However, at long last, each single data will be used in both training and testing, but in a specific way. For instance, in the cross-validation technique, the dataset is split into folds, say k-fold, where k stands for the number of folds or splits. During the training phase, k-1 folds are used to train the algorithm or model, and the remaining split that is held out is used for the testing. This is repeated k times until all the splits are used in both training and testing, but in every single iteration, no element in the testing split is used at the same time in the training splits.

Summary and Discussions

Deep learning emulates and learns about the features that may exist in complex datasets using multiple learning layers of processing units to recognise them as the human brain does. Recently, convolutional neural networks have had a significant impact in the field of visual computing.

In visual applications, the prominent deep learning models utilised are the CNNs. CNN models work quite differently from many other learning models because they are inspired by biological processes for learning information in humans.

The architecture of CNN is composed of convolutional layers which pass several filters over the data, e.g., over an image, to obtain a series of convolved feature maps. They also contain pooling layers to down sample these feature maps. These convolutions and pooling layers are usually terminated by one or more fully connected layers and a Softmax output layer. The fully connected layers are joined to the previous layer as well as the subsequent layers. The final layer, known as the output layer, generally has neurons according to classes, where each class is represented by one neuron. Finally, the last layer typically performs a Softmax, and the class with the highest value gets chosen as predicted by the model.

By way of backpropagation, with an update function, the CNN can learn the filters through the layers, whereas the weights between layers of neurons are the objective function for learning in fully connected layers. Thus, using backpropagation, the weights are updated, and the network is iterated until it produces an accurate output with a given threshold. Typically, several epochs are used to train a CNN, as this leads to a good, generalisable set of learned filters and weights.

Training is normally performed by dividing the data into three subsets, where the training set is used for training, the validation set used during the training to examine how accurate the model is on the unseen sample, and lastly, the testing set is used after the training to verify the performance of the model.

Transfer learning is a technique that takes advantage of an already learned algorithm in one domain and applies the concept to another domain. This is normally utilised when the data to be used to train a new algorithm is scarce. Machine learning works under the assumption that learning of a new task commences from scratch, but this is technically impossible in some domains. For example, in the healthcare sector,the availability of data is highly restricted due to privacy issues. As such, getting hold of enough data, for training a machine learning algorithm and testing it can be very difficult. Therefore, transfer learning provides an avenue to apply the knowledge acquired to fields where enough data is not readily available.

Further Reading

Rastegari, M., V. Ordonez, J. Redmon, and A. Farhadi. Xnor-net, Imagenet classification using binary convolutional neural networks. In Proceedings of the European Conference on Computer Vision (ECCV), pp. 525–542, 2016.

Simonyan, K., and A. Zisserman. Very deep convolutional networks for large-scale image recognition. In Proceedings of the International Conference on Learning Representations, ICLR, 2015.

Srivastava, R. K., K. Greff, and J. Schmidhuber. Training very deep networks. pp. 2377–2385. *In*: Proceedings of the Advances in Neural Information Processing Systems (NIPS), 2015.

Zhang, C., S. Bengio, M. Hardt, B. Recht, and O. Vinyals. Understanding deep learning requires rethinking generalisation. In Proceedings of the International Conference on Learning Representations, ICLR, 2017.

CHAPTER 4

Deep Face Recognition

Face recognition is probably the most exciting and prominent application of deep learning applied in the domain of visual computing. It not only can showcase the power of deep learning but also point out some of the concerns of using a methodology whereby a black-box approach is utilised for solving problems whose solution has real-life consequences.

Computer-based face recognition is still riddled with many challenges compared to the face recognition ability of humans. For a human, seeing someone for a brief moment is usually enough to learn about the face. This is because the brain memorises important details relating to the person. As a matter of fact, it is thought that when a familiar face is presented within a different scenario, the brain compares the 'before' and 'after' images, without the use of any significant new information. Nevertheless, in general, for a machine, the variability of the appearance of a face has a direct effect on face recognition.

There are many machine learning algorithms within computer vision specific to face-related functions. The algorithms are either unsupervised frameworks that are not explicitly programmed or supervised, which are based on the idea that someone can select a portion of data with known labels (based on the knowledge of the operator) or feed it to software as a training set.

Principal Component Analysis (PCA), was introduced in 1905 and is an unsupervised machine learning model widely used for reducing dimensionality and image compression. It is a technique for coding large faces into the "face space". In 1991, Eigenface was applied to construct face recognition algorithms. Eigenfaces extract the most important details known as "eigenvectors" from the faces which correspond to the maximum eigenvalues, representing the variations. Essentially, when it comes to face recognition, the concept of an average face is interesting, and in some ways, it appears to "mimic" human face recognition. Attempts to use the average face as a tool for computer-based face recognition seem to be plentiful in the published literature. Another commonly used technique is the Local Binary Patterns (LBP) and is a simple yet powerful feature for texture classification. This approach divides an image into different regions and extracts features from each region separately, and those features are subsequently used for classification.

In conventional computing, a given algorithm is a group of explicitly programmed commands utilised by a machine to figure out or solve a problem. Machine learning approaches permit machines to train by using input data and utilise statistical analysis to infer values that fall within a particular domain. For this reason, machine learning enables computers to build models from samples of data to render a decision-making process to be automated, based on data inputs. As discussed previously, common examples of machine learning are Convolutional Neural Networks (CNNs), which have been utilised recently in the area of computer vision, especially in face-specific applications. CNNs are supervised models that function to gain knowledge by training and can also be used to represent the most discriminatory features, i.e. feature extraction, face recognition, classification, and segmentation. There are several trained models in the literature, e.g., VGGF, FGG16, VGG19, OverFeat.

Thus, before deep learning, most of the face recognition—or for that matter, general image analysis—have utilised one or two layer-based image processing techniques such as filtering, histograms, and feature coding. There was a significant drawback in such methods, mainly because these methods can only solve one aspect of the face recognition problem at the expense of the others. For example, while a Gabo filtering method can enhance face recognition under varying lighting conditions, that same method can perform poorly in the presence of facial expressions and pose variations. As a result, researchers struggled to come up with a coherent and integral method that can solve the bulk of the issues in face recognition efficiently.

However, much of that changed back in 2012, when it was decisively demonstrated that deep learning could take care of much of the problems faced by computer-assisted face recognition at the time. That was when the researchers, through the use of the deep learning model AlexNet, demonstrated that it was consistently ahead of the game on face recognition by winning the ImageNet competition. Since then, it has been an upward trajectory for deep learning where it has stayed ahead of the curve in showing that it can match human performance in face recognition. For example, in 2014, DeepFace has shown that on the Labelled Faces in the Wild (LFW) dataset, it can achieve human-level accuracy, i.e., 97.35% for DeepFace vs 97.533% for humans.

Components of Face Recognition

There are essentially three parts to a modern deep learning-based face recognition system. They are face detection by which a face is successfully identified within the image, face processing by which the face is cropped and often normalised, and face recognition by which a deep learning algorithm is used to classify or match the face.

Though deep learning can be efficiently utilised to solve most of the object recognition problems, for face recognition problems, we still have to resort to a face processing step to extract the face from the scene to ensure the challenge of the facial pose, illumination, facial expressions, and occlusion are minimised.

The prominent part of any deep learning-based face recognition system is the deep features that are derived from a trained CNN model. Several model architectures are available for a user to choose from. These include the AlexNet, GoogleNet, ResNet, and VGG.

Thus, in symbolic terms, a face recognition system can be described as,

$$M[F(I_i), F(I_j)], \qquad (4.1)$$

where I_i and I_j are two face images there compared, F is the deep features from the CNN model, and M defines the matching criteria. Following, usually a very long process of, training with massive datasets and also with the supervision of appropriate loss functions, the optimal layer from which features are to be extracted and compared are determined. The matching process itself can be undertaken using distance measures such as Euclidean distance, Cosine distance, and Support Vector Machines (SVMs).

Thus, today, most of the face recognition in practice is undertaken with the aid of a deep learning model, and as mentioned earlier, there are many to choose from. In what follows, we discuss some examples to further explain the process of deep learning-based face recognition and the challenges one needs to be vigilant about.

Face Image Datasets

There are several face datasets one can utilise for training and testing the face recognition models. Here we provide details of some of them.

The LFW Dataset

The Labelled Faces in the Wild (LFW) is a large dataset of face pictures, which is designed for testing the capability of face recognition in simulated uncontrolled scenarios. All the images have been collected from the

Internet and consist of a spectrum of variations in expression, pose, age, illumination, and resolution. The LFW database contains images of 5749 subjects with a combined total of 13000 images. The images themselves in the dataset have variable and significant background clutter.

The YouTube Faces DB

The YouTube Faces DB, is composed of face videos with varying lighting, pose and age conditions. The database is specifically designed to study and analyse face recognition algorithms in videos. This dataset contains over 3000 videos of over 1500 individuals. The videos have been downloaded from YouTube. The database has been put together by closely following the LFW as a benchmark.

The FEI Dataset

This database contains 200 images of Brazilian students and staff of an equal number of males and females. For each subject, there are 14 images are totalling the number of images in the dataset at 2800. The resolution of the images is 640 pixels by 480 pixels, and all the images are in colour taken against a homogeneous white background. The subjects are between the ages of 19 to 40 years old. And the dataset contains images displaying variations in facial expressions and pose.

The FaceNet Model

The FaceNet model is a pre-trained deep learning architecture inspired by GoogLeNet models for efficient face recognition. For the task of face recognition, it uses an impression of a list of people in a dataset along with data from a new person or people to be recognised. A key element of the FaceNet architecture is the generation of an embedding of a given dimension from a face image of a predefined size. The input image is fed through a deep CNN architecture which has a fully connected layer at the end. This results in an embedding of 128 features that may or may not be visually understandable to a human. Then, for the recognition task, the network can calculate the distance between the individual features of each of the embeddings. Metrics such as the squared error or the absolute error can be utilised to compute the distance between the embedding.

Common FaceNet models use two types of architectures. They are the Zeiler and Fergus architecture and GoogLeNet style Inception model. The essential idea in the training of the FaceNet is the "triplet loss" to capture the similarities and differences between classes of faces in a 128-dimensional embedding. Hence, given an embedding $E(x)$ from an image to a feature set R^n, FaceNet looks into the squared L_2 distance between face images such that this value is small for images of the same identity and is large for the different identities.

Figure 4.1 shows the general architecture of the FaceNet model. An important element of the model is the triplet loss function. Often in common deep learning models, the loss function tries to map all faces of the same identity to a single point in R^n. The triplet loss function attempts to discriminate each pair of faces from one person to all the others, thus enforcing strong discrimination between faces. Thus, the triplet loss function chosen in this model ensures that an image of a specific person is closer to all other images of that person than any other image in the dataset. This idea is illustrated in Figure 4.2. The training process assumes that we pick a random image—the anchor—from the dataset. We would want to ensure that the distance of that image from another image of the same identity—the positive image—is closer to that of the images not belonging to the same person.

The FaceNet model was trained using around 100 million face samples with over 8 million identities. With the optimal embedding dimensionality of the output, the layer is 128, meaning a given face is classified using 128 feature points extracted from it. Experiments were also conducted on varying the training data, and it was reported that after a certain point the number of training samples does not add value to the level of accuracy, i.e., while 10s of millions of samples can improve the accuracy, adding 100s of millions to the sample has diminishing returns in the accuracy of recognition obtained.

Figure 4.1: The general architecture of the FaceNet model.

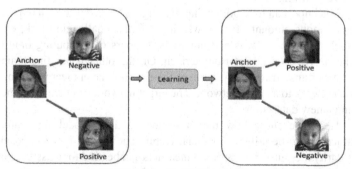

Figure 4.2: Demonstration of how FaceNet is trained using for triplet loss.

Recognition Experiments using FaceNet

One of the main benefits of the FaceNet model is that it can achieve very high classification accuracy using a simpler embedding consisting of only 128 features. Experiments done on faces on both the LFW dataset and the YouTube Faces DB show the recognition is very high, i.e., on the LFW dataset, the recognition accuracy of 98.87% is achieved while on the YouTube Faces DB a recognition accuracy of 95.18% is achieved. It is also crucial to highlight that both the datasets contain faces taken in varying lighting, pose, occlusion, and age conditions. Despite this, the performance on recognition tasks by FaceNet is impressive.

Deep Face Recognition using Partial Faces

Based on the numerous work that has been undertaken in the field of face recognition using deep learning, it is clear that many of the state-of-the art algorithms provide human-level accuracy for face recognition when the query images are full frontal. For example, the FaceNet model, discussed above, provides an impressive level of accuracy for face recognition using frontal facial images.

However, in many practical scenarios, the full face may not be available as a probe or a comparison image in a dataset. Here, we discuss how deep learning-based methods can be taken further forward in that such models can be trained to successfully recognise faces even with partial information. Achieving this can provide the potential for deep learning models to surpass human-level face recognition.

As discussed in the previous chapters, generally speaking, there are several ways one can deploy CNNs today. These include training a network from scratch or fine-tuning an existing model or using off- the-shelf CNN features from a pre-trained model. The latter is referred to as transfer learning.

It is important to highlight that training a CNN from scratch requires an enormous amount of data, which is often a challenging task, e.g., it took millions of faces and hundreds of hours of computing time to train the FaceNet model from scratch. On the other hand, fine-tuning involves transferring the weights of the first few layers learned from a base network to a target network. The target network can then be trained using a new dataset.

Here, we show how the pre-trained VGGF model for feature extraction can be utilised for facial feature coding and how the cosine similarity measure or linear SVM measures can be used for classification for efficient face recognition from partial faces.

The VGG-Face Model

As mentioned in the previous chapter, one of the most popular and widely used pre-trained CNN models in face recognition is the VGG-F model, which was developed by the Oxford Visual Geometry Group. This model has been trained on a large dataset of 2.6 million face images of more than 2.6 thousand individuals. The architecture of VGG-F consists of 38 layers starting from the input layer up to the output layer. As a fixed criterion, the input should be a colour image of 224 by 224 dimensions, and as the pre-processing step, an average is normally computed from the input image.

In general, the VGG-F contains thirteen convolutional layers, each layer having a special set of hybrid parameters. Each group of convolutional layers contains 5 Max-Pooling layers and 15 Rectified Linear Units (ReLUs). After which, there are three fully connected layers (FC), namely FC6, FC7, and FC8. The first two have 4096 channels, while FC8 has 2622 channels, and are used to classify the 2622 identities. The last layer is the classifier which is a Softmax layer whose function is to classify an image. The architecture of the VGG-F is represented in Figure 4.3.

Figure 4.3: An illustration of the basic architecture of the VGG-Face model.

Feature Extraction

Given an input image, X_0, it can be represented as a tensor $X_0 \in R^{HWD}$, where H is the image height, W is the width and D represents the colour channels. A pre-trained layer L of the CNN can be expressed as a series of functions, $g_L = f_1 \rightarrow f_2 \rightarrow \cdots \rightarrow f_L$.

Let $X_1, X_1, ..., X_n$ be the outputs of each layer in the network. Then, the output of the ith intermediate layer can be computed from the function f_i and the learned weights w_i such that $X_i = f_i(X_{i-1}: w_i)$.

As we know, CNNs learn features through the training stage and use such features to classify images later. Each convolutional (conv) layer learns different features. For example, one layer may learn about entities such as edges and colours of an image while further complex features may be learnt in the deeper layers. For example, a result of the conv layer involves numerous 2D arrays which are called channels. In the VGG-F, there are 37 layers, 13 of which are convolutions, and the remaining

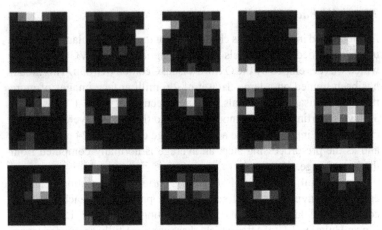

Figure 4.4: An example of the features retrieved from the conv5_3 layer of the VGG-Face model.

layers are mixed between the ReLU, pooling, fully connected, and the softmax. However, after applying the conv5_3 layer to an input image, which has 512 filters with size 3 × 3, the features can be extracted for a classification purpose. By examining the activations of that layer, one can obtain the main features, as shown in Figure 4.4, where a sample of the features is presented.

To decide the best layer within the VGG-F model to utilise for facial feature extractions, one must conduct several trial and error experiments. We tested layers 34 through to 37. In our experiments, we tried other layers. However, the best results are derived from layer 34. It is noteworthy that this layer is the fully connected layer and is placed at the end of the CNN, which means the extracted features represent the whole face.

The features from layer 34 are the results that arise from the fully connected layer FC7 after applying 'ReLU6', which gives a vector of 4096 dimensions. The reason for suggesting layer 34 is the best layer, is inferred as a result of undertaking many face recognition tests where we used the full frontal-face image for both training and testing, and obtained a recognition rate of recognition 100%. The entire process of training and testing through feature extraction is described further in Figure 4.5.

Feature Classification

One of the objectives of the classification is to build a brief model of the distribution of class labels in terms of the predicted features. There are

Input: Training set M, with m classes
n_j = number of images in a given class
for $i = 1$ *to* m **do**
    for $j = 1$ *to* n_j **do**
        im → read an image
        im → resize(im)
        im → normalize(im)
        im features → ExtractFeatures(CNNs(im))
    end for
end for

Figure 4.5: An algorithm explaining how features are extracted from the face data

several techniques for the classification, namely decision trees, k-nearest neighbours (kNN), and SVM.

The SVM is a supervised machine learning algorithm, which can be used for both binary classification and multi-classification problems. The SVM focuses on identifying the "margins" via hyperplanes to separate the data into classes. Maximising the margin reduces the upper bound on the expected generalisation error by creating the largest possible distance between the separating hyperplanes. It is clear that the SVM is geared to solve binary classification problems. In all our experiments, described here, we use the linear SVM to solve the multi-classes classification problem based on One-vs-One (OVO) approach. This is also known as pairwise classification. The OVO decomposition constructs $n(n-1)/2$ binary classifiers for a given n number of classes. Then, for a final decision, the Error Correcting Codes (ECC) combination approach decides how the various classifiers can be combined.

Consider that we have a training dataset (x_i, y_i), we can use the linear SVM such that,

$$\min_{w \in R^d} \frac{1}{2} \|w\|^2 + C \sum_i^N \max(0, 1 - y_i w^T x_i), \qquad (4.2)$$

where, w is a weight vector, N is a number of classes and C is a trade-off parameter between the error and the margin.

Furthermore, one can also utilise the Cosine Similarity (CS) for classification. The cosine similarity is a measure between two non-zero vectors. It uses the inner product space to measure the cosine of the angle between those two vectors. The Euclidean dot product formula can be used to compute the cosine similarity such that,

$$a \cdot b = \|a\| \|b\| \cos \theta, \qquad (4.3)$$

where, a and b are two vectors, and θ is an angle between them.

By using the magnitude or length, which is the same as the Euclidean norm or the Euclidean length of vector $x = [x_1, x_2, x_3, \ldots, x_n]$, the similarity S is computed using the formulation given such that,

$$\| x \| = \sqrt{x_1^2 + x_2^2 + x_3^2 + \mathrm{L} + x_n^2}, \tag{4.4}$$

$$S = \cos \theta = \frac{A.B}{\| A \| \| B \|},$$
$$= \frac{\sum_{i=1}^n A_i B_i}{\sqrt{\sum_{i=1}^n A_i^2} \sqrt{\sum_{i=1}^n B_i^2}}, \tag{4.5}$$

where, A and B are two vectors.

For classification one can compute the CS to find the minimum "distance" between the test image, $test_{im}$ and training images, $training_{im}^n$ by using the Equations 4.6 and 4.7, such that,

$$M_{CS} = min \, (dist(test_{im}, training_{im}^n)), \tag{4.6}$$

where, im is an image number, and n is total images in the training set and,

$$CS(test_{im}, training_{im}^n) = \frac{\sum_{j=1}^m training_{jm}^i test_{im}}{\sqrt{\sum_{j=1}^m training_{jm}^{i2}} \sqrt{\sum_{j=1}^m test_{im}^2}}, \tag{4.7}$$

where, m is the length of the vector.

Experiments

Using the above formulation, some experiments are discussed here showing how the VGG-F behaves and performs on face recognition tasks with partial faces. To undertake this work, we have utilised face images from two popular face datasets, namely, the FEI and LFW. All images from both of the databases were cropped to remove the background as much as possible using a cascade object detector to extract the face and the internal facial features. However, for some images with very complex backgrounds (as in the case of the LFW database), we cropped the faces from those images manually. For the purpose of training, 70% of the images per subject were utilised and augmented through operations such as padding and flipping. The remaining 30% of the images were used for testing in each case.

Experiment 1

In this experiment, using the FEI dataset, ten test sets were generated for each part of the face. Hence there are 600 images per dataset. The parts

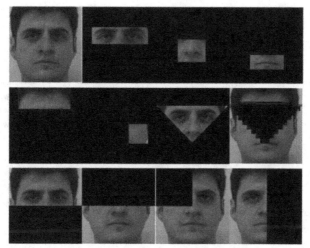

Figure 4.6: A sample of parts of the face for a typical face from the FEI dataset.

of the face chosen were the eyes, nose, right cheek, mouth, and forehead. In addition, faces were generated just with eyes and nose, bottom half face, top half face, right half and three quarters face, and the full-face as well. Figure 4.6 shows a sample of each image.

After extracting the features from all the testing databases by the VGG-F model, the SVM was applied to investigate a rate of recognition in each part separately. As can be seen in the graph shown in Figure 4.7, the highest percentage was with the full face, which was 100% and three quarters as well. At the bottom half, the proportion is decreased gradually and reaches about 50% and this decline is continued until nearly 0.5% at the right cheek.

Experiment 2

This experiment aimed to find the effect of the rotation of the face on recognition. Here 18 separate datasets were generated, and each one had 600 images. These datasets were named R10, R20, R30… R180, etc. The symbol R indicates the rotation in degrees applied. Figure 4.8 shows some samples of rotated faces.

As in the previous experiment, the features that were extracted from all the images were by using the VGG-F model, and these features were tested by SVMs. The best angles that produced the optimum recognition were 10°, 20°, and 30°, the recognition rate at these degrees are 100%, 100%, and about 97% respectively. On the other hand, at 50° the recognition rate dropped down gradually to approximately 27% and this

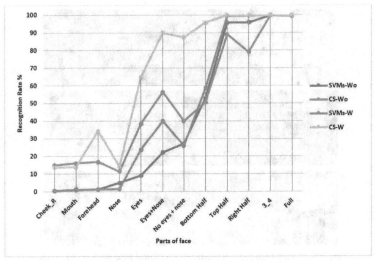

Figure 4.7: Face recognition rates (%) for partial faces, for the FEI face data.

Figure 4.8: Sample of rotated faces from 10° to 180°.

decline continued to decrease until it reached 1% at the inverted face. Figure 4.9 presents the rate of recognition at each angle.

Experiment 3

In this experiment, nine test sets were used for zoomed faces (zoom 10, zoom 20, zoom 30, ..., zoom 90). In addition, each test set had 600 images and all of them were tested individually to find out which degree of zooming can a machine fail in the recognition and which degree had a good rate of recognition. Figure 4.10 illustrates the samples of zoomed faces.

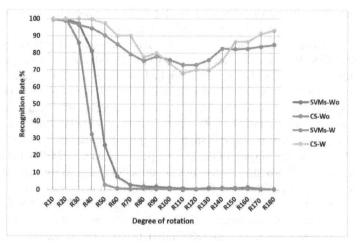

Figure 4.9: Face recognition rates (%) for partial faces with rotation for data from the FEI database.

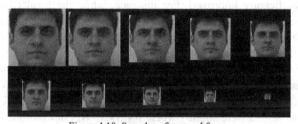

Figure 4.10: Samples of zoomed faces.

After extracting the features for all the faces by using the VGG-F model and linear SVM that was used to classify all of the faces, the following results were formed; the highest recognition rate is at zooming 10°, 20° and 30° respectively which is 100% at each of them. This percentage is slightly decreased at the zooming angle of 50°, which is about 97%. On the other hand, at 60° the recognition dropped down gradually to roughly 46%. Contrary to that the recognition proportion reached near 0.5% at zooming 70, 80, and 90.

Experiment 4

In this experiment, twelve test sets were generated with each test corresponding to a specific part of the face, from faces in the FEI dataset. This time both the full and parts of the face were used in training. The parts of the face that were tested included the eyes, nose, right cheek, mouth, and forehead. Also, faces were generated using just the eyes and nose, the bottom half of the face, the top half of the face, the right half, and a three-quarter of the face, in addition to the full face.

After extracting features from the VGG-Face model, the CS without parts (CS-Wo) and the linear SVM (with 19900 binary classifiers) without parts (SVM-Wo) were applied to investigate the rate of recognition for each facial part separately. The highest recognition rate is achieved when using the full face, and the three-quarter of the face since a recognition rate of 100% is achieved when using both classifiers. However, the recognition rate starts to slightly decrease when using the right half and the top half of the face, respectively with SVM-Wo, but in the case of CS-Wo, the rate remains at 100%. Similarly, as we approach the bottom half of the face, the rate of recognition decreases further to approximately 50% in the case of SVM-Wo and about 60% for CS-Wo. This decline continues until nearly 0.5% for the right cheek.

To measure the rate of recognition using parts of the face, we repeated the above procedure, but this time we added the individual parts of the face into the training set also. The recognition rates in this case significantly improved. For instance, while the results from the right cheek previously were nearly 0%, the recognition rate increased to 15% when using both the classifiers. Also, in the case of the combined eyes and nose features, recognition was previously at 22% for SVM-W and 40% for CS-W. However, in this case, recognition improves to approximately 57% for SVM-W and 90% when using CS-W. We have noticed, however, that not all recognition rates steadily increase. In some cases, the results were slightly worse for SVM-W. For example, a slight decrease in the recognition rate was observed when the bottom half of the face was tested, a recognition rate of 53% was achieved but decreased to 51%. In contrast, the CS-W approach has produced a significant improvement; for example, the recognition rate for combined eyes and nose increased from 40% to 90%.

Experiment 5

In this experiment, twelve test sets were generated with each test corresponding to a specific part of the face, from faces in the LFW dataset—again using both the full and parts of the face were used in training. The parts of the face that were tested included the eyes, nose, right cheek, mouth, and forehead. Also, faces were generated using just the eyes and nose, the bottom half of the face, the top half of the face, the right half, and a three-quarter of the face, in addition to the full face.

All the extracted features from the VGGF model were passed onto both the classifiers (SVM and CS), in both experimental conditions, namely without parts in training (SVM-Wo and CS-Wo) and with parts in training (SVM-W and CS-W). To investigate the recognition rates for each facial part, each classifier was applied separately. In the case of without facial parts, it is clear that in a general CS-Wo outperforms the SVM-Wo for most of the regions of the face. The recognition rates for the right cheek, mouth, forehead, and nose are low, with about 1% for both the classifiers. In contrast, the rate of recognition increases significantly for the facial parts such as the eyes, which touches 40% using CS-Wo. We noticed that as we increased the proportion of the face, the recognition rate also improved significantly, with the best recognition rate of 100% for the ¾ face and full face. Again, in all tested cases, we note that CS outperforms the SVM.

Similar to the previous experiments, we repeated all the tests using facial parts and the same classifiers, but this time we added the facial parts to the training sets (SVM-W and CS-W). The results, as shown in Figure 4.11, indicate a marked improvement in the recognition rates when using SVM-W. For instance, the rate of recognition for the right cheek was about 1% and now reached almost 10%. When considering the mouth, forehead, and nose, we also observe slight improvements in the recognition rates for both classifiers. As we increase the proportion of the face, the recognition rates also significantly improve with CS-W, which reaches 70% instead of about 42% for the eyes. However, this improvement did not occur when using SVM-W (recognition rate of 7%). Moreover, for the images with occluded regions of eyes and nose, bottom, top and right half, the rate of recognition enhanced significantly

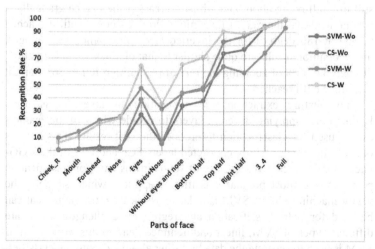

Figure 4.11: Recognition rates (%) using images from the LFW dataset. based on parts of the face using two classifiers: SVM and CS. The tested stimuli include conditions with and without individual facial parts during the training phase.

with CS-W but slightly decreased for the ¾ of face, from approximately 94% to 93.5%.

Discussions

In this chapter, we have discussed how deep learning can be used for face recognition. In particular, we have discussed how the FaceNet model can be used for recognition when information from the full face is available. We have also shown how the VGG-Face model for extracting the features and support vector machines, SVMs for classification, for face recognition on partial faces perform. The ability for existing machine-based face recognition algorithms to perform adequately in cases of partial facial data such as occluded faces, rotated faces, or zoomed out faces, as cues, remains a challenging task in the field of visual computing. However, the experiments presented demonstrate how deep learning can be utilised for face recognition using partial facial cues. Thus, given some partial facial data, we have shown how feature extraction can be performed using popular CNNs such as the VGG-F model. We also have shown how classifiers based on popular SVMs and CS can be applied to undertake facial recognition tasks.

The results using the VGG-Face model on partial faces showed that the small parts of the face have a low rate, especially cheeks and forehead. In the eyes, the recognition rate is slightly improved, but in both eyes and nose, the recognition was better. There is a sharp improvement from the start of the half bottom. For the rotation, in the ten degrees, the machine still had full recognition, and whilst increasing the degree gradually it was noticed that the rate of recognition dropped down until it reached 0.3333 at 180 degrees for the inverted face. On zoomed face images, the recognition results are quite similar to that obtained on face rotation, through the 60% of zooming the rate of recognition fell to 36%, and it continued until reached 0.5%.

For feature extraction, the application of layer 34 is used. Furthermore, applying different layers to decide which of the features can be used for classification to produce the best results. This layer is between the first two fully connected layers FC6 and FC7 of the VGG-Face model, which gives a vector of 4096 dimensions. For classification purposes, the most popular classifier is applied, which supports the vector machine SVMs. SVM is a data classification technique that can be used for both classification and regression challenges. There are different types of SVM, linear and nonlinear, and in this work, a linear SVM is used to separate the data by fitting all images into the classifier.

In this chapter, we have discussed a set of experiments relating to face recognition using partial facial data. Our results show that as the proportion of the face gets small regardless of the prominent nature of the facial features such as the eyes, nose, or mouth, the rate of recognition is poor. However, we note that even in the case of machine-based face recognition. The eyes appear to carry more recognition cues compared to other individual facial features. Furthermore, when it comes to rotated faces, we note that it would be better to avoid highly rotated faces (e.g., faces rotated between and 110° and 120°) as they appear to perform very poorly in recognition tasks regardless of incorporating rotated faces into the training data. In the case of zoomed-out faces, again it is advisable not to use highly zoomed out faces (e.g., faces zoomed-out to 70% to 90%) as probes. Finally, we note that the CS measure greatly improves the performance of the classification when compared to both the linear and kernel SVMs.

Further Reading

Elmahmudi, A., and H. Ugail. Deep face recognition using imperfect facial data. Future Generation Computer Systems 19: 213–225, 2019.

Elmahmudi, A., and H. Ugail. The Biharmonic Eigenface, Signal, Image and Video Processing 13(8): 1639–1647, 2019.

He, K., X. Zhang, S. Ren, and J. Sun. Deep residual learning for image recognition. In Proceedings of the IEEE conference on computer vision and pattern recognition, pp. 770–778, 2016.

Hu, J., L. Shen, and G. Sun. Squeeze-and-excitation networks. In Proceedings of the IEEE Conference on Computer Vision and Pattern Recognition pp. 7132–7141, 2017.

Huang, G. B., M. Ramesh, T. Berg, and E. Learned-Miller. Labeled faces in the wild: A database for studying face recognition in unconstrained environments. Technical Report 07–49, University of Massachusetts, Amherst, Tech. Rep., 2007.

Krizhevsky, A., I. Sutskever, and G. E. Hinton. Imagenet classification with deep convolutional neural networks. In Advances in Neural Information Processing Systems, pp. 1097–1105, 2012.

Schroff, F., D. Kalenichenko, and J. Philbin. FaceNet: A Unified Embedding for Face Recognition and Clustering, 2015 IEEE Conference on Computer Vision and Pattern Recognition (CVPR) 2015.

Simonyan, K., and A. Zisserman. Very deep convolutional networks for large-scale image recognition, arXiv preprint arXiv: 1409.1556, 2014.

Szegedy, C., W. Liu, Y. Jia, P. Sermanet, S. Reed, D. Anguelov, D. Erhan, V. Vanhoucke, and A. Rabinovich. Going deeper with convolutions. In Proceedings of the IEEE Conference on Computer Vision and Pattern Recognition, pp. 1–9, 2015.

Taigman, Y., M. Yang, M. Ranzato, and L. Wolf. Deepface: Closing the gap to human-level performance in face verification. In Proceedings of the IEEE Conference on Computer vision and Pattern Recognition, pp. 1701–1708, 2014.

Ugail, H. Secrets of a smile? Your gender and perhaps your biometric identity, Biometric Technology Today, 2018(6): 5–7, 2018.

Ugail, H. and A. Al-dahoud. Is gender encoded in the smile? A computational framework for the analysis of the smile driven dynamic face for gender recognition. The Visual Computer, 34(9): 1243–1254, 2018.

CHAPTER 5

Age Estimation from Face Images using Deep Learning

A person's age has always been an important attribute of identity. Equally, age also has been an important factor for social interaction. The human face and the physical attributes serve as important elements that facilitate the prediction of a person's age. Thus, age estimation from the face by numerical analysis finds many potential applications such as the development of intelligent human-machine interfaces and improvement of safety and protection in various sectors such as transport, security, and medicine. In this chapter, we discuss how computer-based age estimation can be performed. More specifically, we look into how deep learning-based age estimation from facial images of individuals can be undertaken. We show how a transfer learning approach based on existing pre-trained deep learning models can be utilised for feature extraction and classification from facial images to accurately predict the age of individuals. Details of the deep learning methodology, experimental setup, and the results are discussed in this chapter.

Introduction

In recent decades, with the growing need to automate recognition and surveillance systems, analysing the human face has become an important topic of interest to the field of visual computing. Examples of facial analysis include detection, face recognition, gender classification, facial expression recognition, and equally important, facial age estimation. Being able to predict someone's age from a photo automatically presents crucial advantages in many application areas. These include age-based access control, age estimation in crime investigation, age adaptive targeted marketing, customer relationship management, development of recommendation systems for entertainment, soft biometrics, targeted advertising, intelligent surveillance, internet access control, ethnicity classification, and social media analysis.

For example, having access to a tobacco vending machine is a great convenience for people who smoke. However, providing access to the vending machine will come at a cost in that the ability for underage individuals to buy tobacco products from the vending machine must be restricted. Similarly, this principle applies to many other areas, such as providing access to restricted areas and websites based on the person's age. Furthermore, an automatic age estimation tool would be of great help for aiding with the task of information retrieval. The Internet, for example, with billions upon billions of photos, is recognised as the world's biggest image database. Despite the abundance of visual resources, looking around the world-wide-web for images is not a trivial task. Automatic image retrieval plays a role that is considerable in managing social networking sites for many crucial tasks such as tagging individuals and estimating age groups. Aided by the aid of efficient age estimation, image retrieval from the Internet, for example, can be made more effortless and smarter.

Similarly, many systems currently exist for age progression, for example, to assist with finding missing people. The precision of age synthesis algorithms is usually evaluated according to two elements; the degree to which it keeps the identity of the subject and its capability to make a genuine face that meets the projected age. The most appropriate way of assessing the latter is via age estimation while, the former, is assessed using an appropriate image similarity measure. Moreover, since age progression algorithms are data-driven, an accurate age estimation algorithm may be useful to crawl large datasets for photos to help find missing people.

Thus, in general, the task of automatic estimation of age by an algorithm is useful in applications where the objective is to determine the age of an individual without prior knowledge. However, estimating age from facial images is rather a complex task. To do this, with the aid of an algorithm, we can consider a person of a given age to belong to a certain class among several classes. This problem is much more complicated when compared to face detection or gender estimation because the latter is classed as a problem involving binary classification. This problem is further complicated by the fact that people of various ethnicities, gender, and lifestyles age very differently. Thus, a person's living condition, lifestyle, and quality of health all play a vital role in defining the person's biological age. Additionally, automatic age estimation from photos presents the already existing challenges in face analysis inbinary classification scenarios such as face detection or gender classification. These include the variation in head pose, occultation, facial expression and lighting conditions.

Thus, during the pre-era of deep learning, researchers categorised features from the face into local, global and hybrid forms. Others utilised image representation mechanisms for age modelling that took advantage of active appearance models (AAM). Such models were extensively used to learn shape and texture features from training images that used entities such as principal component analysis, correlation filters and Random Forests for dimensionality reduction.

Like many visual computing-related problems, there have been many computer-assisted algorithmic solutions proposed for the problem of automatic age estimation from a face photo of an individual. Typically, such methods are two-step based. The first step involves extracting features from the image while during the second, some form of learning would take place to estimate the age. For example, one could utilise anthropometric models and wrinkle patterns of the face for age estimation. Traditional image processing tools such as the AAM can be used to extract and compare prominent features around the eyes, cheeks and forehead. Other methods such as Linear Discriminant Analysis (LDA), kNN (k-Nearest Neighbours), Local Binary Patterns (LBP) and Gabor wavelets, fuzzy logic and Support Vector Machines (SVMs) can also be used for image classification based age estimation. However, like many other image analysis applications using such low dimensional image analysis tools, the final solution is often not scalable and not robust. Hence, researchers have turned to deep learning for a robust solution to the problem of age estimation.

Deep Convolutional Neural Networks for Age Estimation

During the recent past, Deep Convolutional Neural Networks (CNNs) have shown exceptional performance on age estimation tasks from photos. This progress has shown the death of most traditional image processing techniques that have been in use in this field for a very long time. As highlighted earlier, such outdated techniques relied on mostly handcrafted and hardcoded features which are prone to errors and faced issues relating to computational accuracy.

A common mechanism for utilising a CNN for the task of age estimation is to make use of one of the pre-trained models such as the VGG and GoogleNet. Other architectures such as the Directed Acyclic Graph CNN (DAG-CNN) are also utilised. A DAG is simply a feed-forward CNN that consists of convolution layers, Multi-output layers (ReLU), normalisation layers and pooling layers. Usually, the ReLU layer is connected to an average pooling layer, and then it is normalised to feed to a fully-connected layer that computes an inner product with a number of outputs. The output is sent to a suitable Softmax function for final classification.

Popular Datasets

To measure the efficiency of age estimation algorithms, one needs to appropriately train and test the various algorithms for training and testing. The desired quality of images in the datasets must also meet certain criteria, i.e., there should be low variation in conditions such as the post, lighting and texture. The datasets must be representative of gender, ethnicity and age ranges. There should be enough facial images for meaningful training and testing of the developed algorithms. Here we discuss some of the popular datasets that are available for researchers to test and verify their algorithms.

1. The MORPH dataset [1] is the large commonly used facial database for facial age estimation and facial age synthesis research. The dataset comprises two albums. Album 1 contains 1690 face images with age, ethnicity, gender, height and weight of the individual available. Album 2 contains 78,207 images with age ranges of 15 to 80 years with variations in facial expression, illumination and resolution. Arguably, MORPH is the most popular dataset among researchers working in age estimation from faces.

2. The FERET dataset [2] contains 14,126 grayscale face images of 1199 individuals of varying ages. The images themselves have variations in pose, ethnicity and gender. The overall quality of images is generally good as they appear to be collected in controlled environments.

3. The FG-NET dataset [3] contains 1002 colour and grayscale images with a mixture of ages ranging from infants to older people. The data has an adequate level of ethnicity variation and has labelled information with the actual ages of the subjects. The visual quality across the image is not uniform and have various illumination, pose and background variation as well as other types of noise.

4. The PAL dataset [4] contains 1,142 face images ranging from 19 to 93 years with 225 males and 350 females. Overall, the images present in this dataset have good resolution.

5. Cross-Age Celebrity dataset [5] includes images of 2000 celebrities collected from the Internet. The dataset has variations in the quality of images and in some case, the ages are probably estimated rather than being the actual ages of the celebrities.

6. Apparent Age Estimation (LAP) dataset [6] consists of 7,591 images of varying ages, ethnicity and gender which have been taken in unconstrained environments. The images themselves were labelled by human observers where statistical mean and standard deviation of the assessment by the observes are used to label the images.

Performance Measures

The performance of an age estimation algorithm can be measured using a number of techniques. Common techniques used for this purpose are, Mean Absolute Error (MAE) and the cumulative score (CS). MAE is defined to be the summation of errors between the predicted age and the actual age divided by the number of testing images in the dataset, i.e.,

$$\text{MAE} = \frac{1}{N}\sum_{k=0}^{n} |s_1{}^k - s_2{}^k|, \qquad (5.1)$$

where N refers to the number of testing samples, s_1 refers to the actual age of a sample k and s_2 refers to the estimated age of the same sample.

The cumulative score (CS) is derived from MEA, which computes the proportion of the test images that have an absolute error less than the specified number of years. Thus, CS is defined as,

$$\text{CS} = \frac{n^L}{N} \times 100\%, \qquad (5.2)$$

where n^L refers to the number of testing samples that have errors less than a specific number of years L.

Age Estimation using the VGG Model

As discussed earlier, the VGG-16 is a publicly available model that was trained using 2.6 million face images of 2622 unique subjects. The model is configured to take a fixed-sized [224 × 224 × 3] RGB image as an input, as a form of preprocessing whereby all the images used are centre-normalised. The network is made of a stack of 13 convolutional layers with filters having a uniform receptive field of size 3 × 3 and a fixed convolution stride of 1 pixel. These convolution layers are followed by five max-pooling layers. Finally, the CONV layers are then followed by three fully connected layers; FC6, FC7, and FC8. The first two layers have 4096 channels, while the FC8 has 2622 channels which can be used to classify the 2622 identities.

Feature Extraction

For face representation weights from different layers of the VGG-Face model are used to extract deep features. Dimensions of the resulting features are then reduced before using a regression for age estimation.

Given an input image X_0 represented as a tensor $X_0 \in \mathbb{R}^{H \times W \times D}$ where H is the image height, W is the width and D the colour channels, and a pre-trained L layered ConvNet expressed as a series of functions $q_L = p_1 \rightarrow p_2 \rightarrow \cdots p_L$. To fully investigate and evaluate which layer yields the best age descriptor, the activation of five layers; the last two convolution layers (conv5_2, conv5_3), the last max-pool layer (pool5) and the first two fully connected layers FC6 and FC7 of the VGG-Face model are used as separate feature channels. The choice of layers has been restricted to the top 5 layers, because going further down yields extremely huge dimensions that will result in no significant gains even after reducing the dimension.

Dimensionality Reduction and Regression

Due to the large dimensions of the extracted features, ranging from 4097 in FC7 to 100352 in conv5_2, there is a need to reduce the feature size, thus removing redundant information. Moreover, it is a well-known fact that, for n observations and p features, the regression estimate is not well-defined in a situation where $p > n$. It is common for a researcher to use the PCA for dimensionality reduction. However, there are obvious problems with PCA, and a better alternative often is the partial least squares regression (PLS) that can be simultaneously applied to reduce the dimension and regress. Hence, the relationship between the extracted features X and the vector of ages Y can be formulated as,

$$Y = X\beta^{PLS} + b, \tag{5.3}$$

where b is the intercept.

Age Estimation

To estimate the age, features extracted using different layers of the VVG model are compared in order to identify which weights of the deep network carry the most optimal ageing information. To enhance the specificity of the technique, all images can be first cropped to a size of 224×224, and then a data preprocessing step can be deployed. Finally, the extracted features are fed can be fed into a partial least square algorithm (PLS) for age prediction.

Experiments using the FGNET dataset

Given a set of images from the FGNET dataset, they can be aligned using landmark annotations. Furthermore, their backgrounds can be removed to increase image purity. Thereafter, data augmentation can be conducted. This is a popular technique used to increase data size during the training phase. As a result of augmentation, each image was responsible for the generation of 7 additional images achieved via random cropping and warping to the mean shape, as shown in Figure 5.1.

By utilising the procedure describe in [5], five sets of estimations can be conducted. Each estimation can be performed by extracting features using one of the five layers of the VGG-Face model, namely, conv5_2, conv5_3, pool5, FC6, or FC7 layers.

To evaluate the performance of the estimation procedure, the two metrics, MAE and CS can be used. Comparison of the performance of the five CNN features represented in Table 5.1 shows that conv5_2 activations give the most minimal estimation error. It is also obvious that the performance degrades as one moves higher along the hierarchy. This suggests that generic features learnt from intermediate layer activations carry more ageing information than the latter layers that are more specific to the problem of face identification. The dimensionality reduction capability of PLS is also remarkable, as it reduces thousands of features to just 18 latent variables.

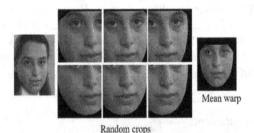

Mean warp

Random crops

Figure 5.1: Image Pre-processing pipeline.

Experiments using the FGNET Dataset

Since some of the recent works on estimation use Morph album II rather than FGNET, it is then ideal to evaluate the proposed algorithm using the Morph II dataset.

In this second experiment, two sets of tests were conducted, by utilising aligned and unaligned images, denoted as *wAlg* and *woAlg* respectively. Besides the above-mentioned preprocessing omissions, all other steps used in the first experiment for feature extraction

Table 5.1: Evaluation of features extracted from different VGG layers.

Layer	Latent Variables	MAE	CS < 10 years
conv5_2	18	2.70	100%
conv5_3	18	2.83	99.01%
pool5	18	2.97	98.05%
FC6	18	3.89	96.31%
FC7	18	5.51	84.21%

Table 5.2: Evaluation of the results using Morph II.

Layer	Tr. Set	Latent Vars.	MAE	Avg. MAE	CS < 10 years
conv5_2	S1	17	3.93	3.92	96.71%
	S2	17	3.91		
conv5_3	S1	17	3.95	3.94	96.61%
	S2	17	3.93		
pool5	S1	17	4.06	4.05	96.06%
	S2	17	4.03		
FC6	S1	24	4.33	4.31	94.32%
	S2	24	4.29		
FC7	S1	24	4.50	4.51	93.26%
	S2	24	4.51		

Table 5.3: Evaluation using Morph II.

Layer	Tr. Set	Latent Vars.	MAE	Avg. MAE	CS < 10 years
conv5_2	S1	17	3.84	3.83	96.82%
	S2	17	3.82		
conv5_3	S1	17	3.87	3.87	96.75%
	S2	17	3.86		
pool5	S1	17	4.01	3.99	96.18%
	S2	17	3.97		
FC6	S1	24	4.27	4.26	94.43%
	S2	24	4.25		
FC7	S1	24	4.45	4.45	93.40%
	S2	24	4.45		

and estimation were repeated. Eventually, various estimations were conducted using the same five CNN activations. Comparison of the estimation results presented in Table 5.2 and Table 5.3 corroborate the findings of the first experiment. Once again, conv5_2 activations give superior performance. The results also show that image alignment increases the performance of the technique.

Further Examples of Deep Learning-based Techniques for Age Estimation

In the above, we have discussed how a deep learning model can be executed for age estimation from photos. This is still very much a ripe area of research. There has been much work in this area over the past decade or so. Here we briefly outline some of the other techniques methods for age estimation using deep learning.

A notable piece of work in this area was carried out by Dong et al. They proposed an end-to-end CNN system for age estimation from face images [7]. The method achieved an average MAE value of 3.63 years on the MORPH2 dataset. Their CNN model estimated the age from an image directly by a deep learned end-to-end system. All the model parameters were learned instead of handcrafting them. Similarly, Yang et al. [8] used a 3-layers ScatNet for feature extraction, whereby PCA is used to reduce the feature dimensions. ScatNet, with its deep convolutional network, showed that it could achieve an MAE value of 3.49 on the MORPH dataset. The work of Wang et al. demonstrated a CNN model combined with Support Vector Regression (SVR) could be used for pattern learning for age estimation achieving similar MAE values [9].

Rothe et al. [10], proposed a deep age estimation method called Deep Expectation of apparent age (DEX). The DEX method predicted the estimated age from a photo by classifying the images into several age groups and then using a Softmax to compute the final prediction on the age. DEX achieved an MEA of the value of 3.25 years and 4.63 years on MORPH2 and FG-NET, respectively. Furthermore, DEX achieved an MEA value of 2.68 years and 3.09 years on MORPH2 and FG-NET. In some cases, the CNN model on which DEX is based failed to estimate facial images belonging to individuals with darker tones and those wearing glasses.

Similarly, Niu et al. [11] utilised an Ordinal Ranking-CNN (OR-CNN) method to transform the original problem age estimation into a smaller set of subproblems involving binary classification which was then enhanced with a Softmax classifier. Thus, the necessary training of CNN was done using binary ordinal age labels, one for each age group. Then, the ordinal regression was converted to sub-problems, whereby each of the subproblems was assigned to the input layer of CNN. Experiments undertaken using OR-CNN showed that it is possible to achieve an MEA value of 3.27 years on the MORPH dataset. Also, Wang et al. investigated a new approach to extract features based on deep learning [24]. They used CNN to extract the features and Support

Vector Regression (SVR) to learn ageing patterns. They didn't concern themselves with the influence of gender and ethnicities where they used only the age information. Further, Chang and Chen reduced MEA to 3.74 years on the large dataset MORPH Album 2 and outperformed the compared methods [12]. They proposed an age ranking approach, CSOHR which is based on ordinal regression. They didn't consider the changes of neutral facial expressions which have effects on age estimation.

More recently, Liu et al., a very lightweight framework based on scattering transformation for feature extraction [13]. They then utilised SVM for age classification. This framework is composed of three key steps, namely, (1) classification of gender, (2) detection of an age label based on the gender classification and, (3) determination of the age value from the age label. Experiments based on this method have shown that it can achieve a reduced MAE value of 3.92 years on the FG-NET dataset when compared to previous techniques. Following this work, Taheri et al., through their DAG-VGG-16 and DAG-GoogleNet CNNs showed that classification accuracy of MEA of 2.81 could be achieved on Morph dataset [14]. They utilised multi-scale features of the face and fused them. These features included the use of handcrafted parameters on each of the CNN layers. Similarly, some studies show how additional learning information, such as race and gender into the CNN model can achieve superior results. For example, the method of RAGN [15], showed an ensemble structure with three networks to determine age, gender, and race for feature definition and classification. These techniques, combined with the extreme learning machine, demonstrated that an MEA value of 2.61 years on Morph-II could be achieved.

Discussions

From the material presented in this chapter, it is clear that deep learning has much merit when it comes to estimating age from facial images. Results of extensive experimentation suggest that deep learning can produce human-level accuracy—and in some cases even outperform the human—in accurately guessing the age of a person from a photo.

For example, we have shown that facial feature extraction using weights of pre-trained CNN has merit for age estimation. Using activations from different layers of the VGG-Face model, we can see that both on the FGNET and Morph Album II databases, good results can be obtained. The simultaneous dimensionality reduction capability such as the use of PLS shows that promising results can be achieved without having to train a CNN from scratch. Much of the deep learning-based age estimation results discussed in this chapter clearly show the superiority it's compared to traditional image classification and analysis techniques.

Although superior in its performance, deep learning faces issues and obstacles which require further research and development. Any CNN is usually computationally extensive—especially in the training phase. CNN also requires careful preparation and handling of training data. Popular CNNs such as the VGG and GoogLeNet is extensive in features and often require strategies for dimensionality reduction. For these purposes, techniques such as scattering transform, transfer learning and ordinal ranking learning can be combined with the base CNN models to ensure accuracy and computational efficiency.

References

[1] Ricanek, K., and T. Tesafaye. Morph: A longitudinal image database of normal adult age-progression. Proceedings of IEEE 7th International Conference on Automatic Face and Gesture Recognition IEEE Southampton, pp. 341–345, 2006.

[2] Phillips, P. J., S. Z. Der, P. J. Rauss, and O. Z. Der. FERET (face recognition technology) recognition algorithm development and test results. Adelphi, MD: Army Research Laboratory, 1996.

[3] Crowley, J. L. and T. Cootes. Face and Gesture Recognition Working Group, [online] Available: http://www-prima.inrialpes.fr/FGnet/. 2009.

[4] Minear, M. and D. C. Park. A lifespan database of adult facial stimuli. Behavior Research Methods Instruments & Computers 36(4): 630–633, 2004.

[5] Chen, B. C., C. S. Chen, and W. H. Hsu. Cross-age reference coding for age-invariant face recognition and retrieval. European Conference on Computer Vision, pp. 768–783, 2014.

[6] Escalera, S., X. Baro, H. J. Escalante, and I. Guyon. ChaLearn looking at people: A review of events and resources. Proceedings of the International Joint Conference on Neural Networks (IJCNN), 2017.

[7] Yi, D., L. Zhen, and Z. L. Stan. Age estimation by multi-scale convolutional network. Asian Conference on Computer Vision, 2014.

[8] Yang, H., B. Lin, K. Chang, and C. Chen. Automatic age estimation from face images via deep ranking. Proceedings of BMVC 55: 1–55, 2015.

[9] Wang, X., R. Guo, and C. Kambhamettu. Deeply-learned feature for age estimation. Proceedings of IEEE Winter Conference on Applications of Computer Vision, pp. 534–541, 2015.

[10] Rothe, R., R. Timofte, and L. Van Gool. Deep expectation of real and apparent age from a single image without facial landmarks. International Journal of Computer Vision 126(2–4): 144–157, 2018.

[11] Niu, Z., M. Zhou, L. Wang, X. Gao, and G. Hua. Ordinal regression with multiple output CNN for age estimation, Proceedings of CVPR, pp. 4920–4928, 2016.

[12] Chang, K., and C. Chen. A learning framework for age rank estimation based on face images with scattering transform. IEEE Transactions on Image Processing 24: 785–798, 2015.

[13] Liu, K., and T. Liua. A structure-based human facial age estimation framework under a constrained condition. IEEE Transactions on Image Processing 28: 5187–5200, 2019.

[14] Duan, M., K. Li, and K. Li. An ensemble cnn2elm for age estimation. IEEE Transactions on Information Forensics and Security 13(3): 758–772, 2017.

[15] Taheri, S., and O. Toygar. On the use of DAG-CNN architecture for age estimation with multi-stage features fusion. Neurocomputing 329: 300–310, 2019.

Further Reading

Bukar, A., and H. Ugail. On automatic age estimation from facial profile view. IET Computer Vision 11(8): 650–655, 2017.

Bukar, A., and H. Ugail. A Nonlinear Appearance Model for Age Progression. In Advances in Soft Computing and Machine Learning in Image Processing (Springer), Hassanien, Aboul Ella, Oliva, Diego Alberto (Eds.), Springer, 2017.

Bukar, A., and H. Ugail. Facial age synthesis using sparse partial least squares (the case of Ben Needham). Journal of Forensic Sciences 62(5): 1205–1212, 2017.

Bukar, A., H. Ugail, and D. Connah. Automatic age and gender classification using supervised appearance model. Journal of Electronic Imaging 25(6); 061605, 2016.

Cootes, T. F., E. J. Gareth, and T. J. Christopher. Active appearance models. IEEE Transactions on Pattern Analysis & Machine Intelligence 6: 681–685, 2011.

Yang, S., and D. Ramanan. Multi-scale recognition with DAG-CNNs. Computer Vision (ICCV) 2015 IEEE International Conference, pp. 1215–1223, 2015.

CHAPTER 6

The Nose and Ethnicity

The nose is the most central feature on the face, which is known to exhibit both gender and ethnic differences. It is a robust feature, invariant to expression, and known to contain depth information. In this chapter, we show how deep learning can be utilised for binary ethnicity classification from images of the nose. We show this through a series of experiments using pre-trained deep learning models such as the ResNet-50, ResNet-101, ResNet-152, VGG-Face, VGG-16, and VGG-19, for feature extraction and a linear Support Vector Machine for classification.

Introduction

It is well established that facial morphology shows a great deal of variation across racial and ethnic groups. This is especially true in the case of the upper region of the face, which includes the nose and eyes. In this sense, the nose is one of the most defining and central features. Nasal anthropometry includes several parameters, which describe the shape and position of the nose. An abundance of anthropometric research has been carried out which focuses on the nose as a stand-alone feature. For example, ocular anthropometry describes the process of taking external measurements of the eye region, either directly or from photographs. Such information can then be used to establish racial and ethnic differences in surface anatomy and the surrounding regions of the eye in population studies.

Ethnic variations in facial appearance are significant. In essence, physiological makeup is dependent upon a person's ethnicity. In visual computing, it is well established that facial features are a critical source of information. Facial recognition systems rely on the attributes of the face to aid identification, which is why it is one of the most studied research areas. And, although the face provides demographic information relating to age, gender and, ethnicity, the latter remains the most invariant trait for image-based classification tasks.

It is well known that the nose is a feature that demonstrates significant variation between racial and ethnic groups. The nose is also one of the cardinal features of the face and contributes greatly to facial aesthetics and racial and/or ethnic inclination. In the past, there has been much research demonstrating significant differences in nasal breadth measurements in individuals of different races. For example, researchers investigated the nasal anthropometric features for males and females of Turkish origin [1]. They looked at 115 adults aged 18 to 30 years, whereby a total of 14 nasal parameters were identified and measured. A major finding from their studies is that there is a significant variation in the nasal shape that allows distinguishing not only the gender but the ethnicity itself.

Thus, the human nose is the most centralised and protrusive feature on the face. There are several advantages of using images of the nose as a primary feature for image-based classification. Firstly, the shape of the nose is robust to changes associated with variations in facial expression. It also is unaffected by facial hair. Moreover, anthropometric studies show both gender and ethnicity differences in nose shape and size. The stability of the nose appearance of an individual, and the invariance of the nasal shape over time, highlights the importance of this surface feature for identification.

Over the years, the task of ethnicity classification has been attempted by multiple machine learning algorithms under diverse assignments. For example, using Linear Discriminant Analysis (LDA) one can show that ethnicity classification from a database of 2,630 images from 263 subjects, has over 96% performance accuracy for binary (Asian vs Non-Asian) classification. Similarly, one can use Gabor Wavelet features and Retina Sampling to obtain the most informative features. Here, the researchers used a support vector machine to classify the features into three classes, namely Asian, African, and European. Researchers achieved the following accuracy; 96%, 94% and 93% respectively [2].

Roomi et al. [3] used the Viola-Jones algorithm for face detection after which features such as skin colour, lip colour, and the normalised forehead region were extracted for racial classification. The three races classified as part of the study were Caucasoid, Negroid, and Mongoloid. The authors proposed that the distinction between the three human groups was historically agreed upon by anthropologists and exhibited differences in head shape, skin colour, lips, and hair texture.

For skin colour extraction, the face images were converted from RGB to the YCbCr colour space with each pixel being classed as either 'skin' or 'non-skin.' Feature extraction was carried out by projecting edge maps using Sobel edge detection. On a dataset created from the Yale and FERET image database, the authors reported an accuracy of 81% for the Caucasoid, Negroid, and Mongoloid races.

Yewenberg et al. [4] investigated the classification of ethnicity in addition to traits such as age, gender, and emotion on the Face Attributes Dataset (FAD). The authors conducted baseline experiments using a pre-trained CNN model; AlexNet and compared the results with their novel landmark augmented CNN (LACNN). The tested ethnicity (White) was accurately classified using AlexNet with 82.7% accuracy, whereas LACNN marginally increased the performance accuracy to 83.35%. The research highlighted the strength of CNN architectures and the effectiveness of using a large and varied dataset.

While the task of ethnicity classification has been addressed using images of the entire face. There is a handful of literature, which has investigated the robustness of facial features such as the eyes and the nose, for ethnicity classification. For example, Qiu et al. [5] suggested a model determine ethnicity (Asian vs Non-Asian) from iris images.

Comparably, the nose has also been shown to be an efficient biometric [6]. However, the availability of nose-related literature is limited. Lv et al. [7] introduced a novel 3D Nose Shape Net classifier for gender and ethnicity classification. To construct a 3D Nose Shape Net, a method to return measurements from the nose is required to ascertain the distances between the different noses within a given dataset, and then the images are clustered into groups. By using the nose clustering results, the 3D nose is constructed. Using the Bosphorus3D face image database and the FRGC2.0, the authors reported an ethnicity classification of 89.2% for the Asian ethnicity and 87.4% for the White.

A similar study on the use of 3D nose shape information has also been reported by Drira and colleagues [8]. Song et al. [9] studied extracting the quantity and position of pores on the skin of the nose as a biometric. Performance accuracy of 88.07% was reported, and the research demonstrated that nose-pores are an encouraging avenue for biometric identification.

The feasibility of image-based nose biometrics for determining identity information was investigated by Zehngut et al. [10]. Here we carried two types of experiments. They are, (1) comparing full-face and nose biometrics using the NIST's Face Recognition Grand Challenge (FRGC) database of 12,776 images and (2) comparing fullface and nose biometrics for occluded face images, using the AR Face database of

3,288 images. Discriminative nasal features were extracted using Kernel Class-Dependence Feature Analysis (KCFA) based on Optimal Trade-off Synthetic Discriminant Function (OTSDF) filters.

Thus, the use of nose images, including aspects of the feature such as nose-pores, for demographic (ethnicity) identification has proven successful. The nose is a robust and discriminative feature, and the literature review proves reliability.

In this chapter, the focus is on ethnicity as a demographic trait, visible within the nose, to demonstrate how a series of experiments can be conducted using a range of pre-trained machine learning algorithms which are Residual Neural Network (ResNet) and VGG-Face based. The pre-trained models which are used as part of our proposed methodology are (1) VGG-Face, (2) VGG-16, and (3) VGG-19. The Residual Neural Network (ResNet) used are; (1) ResNet-50, (2) ResNet-101, and (3) ResNet-152.

For the discussions presented here, we define ethnicity as a person's cultural and ancestral background. Images of the nose that were cropped from full-face images were classed as either Pakistani or Non-Pakistani, based on a specific selection criterion, which is discussed in the upcoming feature database section.

The proposed method for ethnicity classification consists of 2 components; (i) Feature extraction using weights of VGG-Face, VGG-16, VGG-19, ResNet-50, ResNet-101, and ResNet-152 and (ii) ethnicity classification using a linear Support Vector Machine (SVM) algorithm as a binary procedure.

Problem Definition

The task of ethnicity classification is identifying to which set of subpopulations a new observation belongs, based on a training set of data. Such a dataset would contain observations whose category membership is known. Consider a dataset of face images F, consisting of N number of face images as given,

$$F = \{f_1, f_2, \ldots, f_{N-1}, f_N\}, \tag{6.1}$$

This is a binary ethnicity classification E_c that aims to differentiate between a given ethnicity E_p and another person not belonging to that ethnicity E_{NP}, as accurately as possible. Suppose, x represents multiple data points or features belonging to a face image. However, within a real-world scenario, input data for ethnicity classification can be based on either full-face or partial face images. In this case, we can explore both full-face x^F and partialface x^{PF} features for ethnicity classification. Both hand-crafted and deep learning features are incorporated into training. This ethnicity classification problem can be defined as,

$$H(x) = \theta_0 + \theta_1 x, \quad x = x^F U x^{PF}, \tag{6.2}$$

where $H(x)$ is the predictor function, θ_0 and θ_1 are constant values. The goal should be to find the perfect values of θ_0 and θ_1 to make the predictor perform as well as possible. Optimising the predictor $H(x)$ is carried out using the training examples. For each of the training examples, we have an input value x_{train}, for which a corresponding output, y, is known in advance. We find the difference between the known and correct value y, and our predicted value $h(x_{train})$.

With enough training examples, these differences provide a useful way to measure the "inaccuracy" of $H(x)$. We can then tweak $H(x)$ by tuning the values of θ_0 and θ_1 to make it "less wrong". This process is repeated until the system has converged on the best values for θ_0 and θ_1.

The Dataset

The Pakistani Face Database has been developed by researchers within the University of Bradford, United Kingdom. 100 Pakistani and 100 Non-Pakistani images were selected from the Pakistani face database. The ethnicity of each image was determined by asking the participants to self-identify their ethnicity. For participants of Pakistani origin, eligibility was dependent on whether both the maternal and paternal parents were of Pakistani ethnicity.

All the facial images were cropped around the region of the nose using Photoshop CS6 to generate a sub-dataset of the varied nose regions, as shown in Figure 6.1. To keep all the relevant information, it was ensured that the entire nose was cropped, clearly showing the nasal bridge, the pronasale (nasal tip) as well as the nasal alars (nostrils).

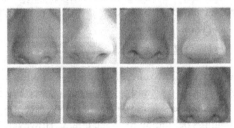

Figure 6.1: An example of nose images that have been processed using Adobe Photoshop CS6. (The images used have been reproduced in line with the approved ethics from the University of Bradford, UK, and participant consent.)

Methodology

Our approach was to extract nose features using six pre-trained machine learning models, three VGG-based and three Deep-learning-based. The models used include; VGG-Face, VGG-16, VGG-19, ResNet-50, ResNet-101 and ResNet-152. Followed by a discriminative binary classification using a Linear Support Vector Machine. A total of eight experiments were conducted on the following datasets: (1) 2,200 isolated nose crops, (2) 1,200 isolated eyes and nose crops, (3) 1,200 isolated nose and mouth crops, and (4) 1,200 isolated mouth and eyes crops.

Data Pre-Processing

All images used in our experiments were resized to 224 × 224 pixels to ensure they conform to the input criteria of the pre-trained models. Furthermore, data augmentation was carried out for the training data using rotations at the degree of 10° right, 10° left, 90°, 180°, and 270° as well as random crops. This led to a minimum of a twelvefold increase in the original data size.

Feature Extraction

The activation of the last pooling layer of each of the ResNet models was used for feature representation.

In total, three sets of features were retrieved using ResNet-50, ResNet-101, and ResNet-152 Neural Networks, per experiment. We decided to exclude the last output layer of the Fully Connected layer (FC), since it was trained on a set of different data (i.e., objects) compared to the facial features which are presently used. Additionally, research proves that the lower layers of the Deep Neural Network are adequate in learning generic features.

Comparably, for the VGG feature representation using the VGG-Face, VGG-16, and VGG-19 models, we used the middle Fully Connected layer (FC7) and discounted FC6 and FC8.

Classification

Having used machine learning-based feature extraction with VGG-Face, VGG-16, VGG-19, ResNet-50, ResNet-101 and ResNet-152 models, independently. A linear classifier was employed for the two-category ethnicity classification.

Support Vector Machines (SVMs) are supervised machine learning models that function to identify a hyperplane, which best classifies data points within a given data space. Published literature has determined that SVMs is a powerful binary classifier that operates by defining an optimum separating hyperplane between two classes of data.

The SVM algorithm classifies data by defining an optimum separating hyperplane between two data classes, which typically involves solving an optimisation problem.

Given a training set,

$$\{(x_i, y_i)\}_{1 \leq i \leq n}, \ x_i \in \mathbb{R}^d, \ y_i \in \{+1, -1\}, \qquad (6.3)$$

SVM finds the optimal hyperplanes by solving,

$$\begin{cases} \min\limits_{w,b} & \dfrac{1}{2} \| w \|^2 \\ \text{with} & y_i.(w.x + b) \geq 1 \end{cases} \qquad (6.4)$$

for $i = 1 \ldots n$ are the observations, where the weights w and b the bias is learned during training. The classifier is learned such that, $y_i \in \{+1, -1\}$, $+1$ denotes images labelled as Pakistani, and -1 denotes non-Pakistani images. A k-fold cross-validation technique was employed to evaluate the performance of the pre-trained models.

Experimental Results

By extracting features from images using three pre-trained deep-learning models; ResNet-50, ResNet-101, and ResNet-152, we achieved results above 90% for the binary classification of a Pakistani nose on a dataset of 2,200 nose images (experiment 1), as shown in Table 6.1.

In contrast, the experiments conducted using the three VGG-based models, namely; VGG-Face, VGG-16, and VGG-19 are varied between each other and are essentially lower when compared to the ResNet models. The performance accuracies are directly comparable to the ResNet models and demonstrate the finding that the accuracy is poorer with VGG than ResNet. This may be attributable to the finding that the ResNet models are classed as very deep learning models, which are architecturally different and are reported to yield higher accuracies.

Table 6.1: Performance accuracy for the binary classification of Pakistani nasal images using residual neural networks.

Feature Extraction Model	Classifier
	Linear Support Vector Machine (SVM)
ResNet-50	*94.1%*
ResNet-101	93.4%
ResNet-152	93.6%

Table 6.2: Performance accuracy for the binary classification of Pakistani nasal images using the VGG model.

Feature Extraction Model	Classifier
	Linear Support Vector Machine (SVM)
VGG-16	*90.8%*
VGG-19	87.5%
VGG-Face	79.9%

However, VGG-16 does outperform VGG-F and VGG-19 in the binary classification task, as shown in Table 6.2.

Additional experiments were carried out using isolated face feature crops, to ascertain whether feature combinations impacted performance accuracy. The feature combinations used consisted of; (1) eyes and nose, (2) nose and mouth, and (3) mouth and eyes, as shown in Table 6.3.

The results achieved from the isolated feature combinations demonstrate that, on average, the features of the nose and mouth yield the highest results (85.63%). Moreover, for all the tested pre-trained models, the nose and mouth feature combination consistently yields the highest performance accuracy.

The results are closely followed by the mouth and eyes and the eyes with nose combination with 82.56% and 81.55% average accuracies, respectively. The results show that the internal features, i.e., eyes, nose, and mouth, provide reliable information to distinguish between the Pakistani and non-Pakistani faces. Importantly, the reported results are in line with literature that proposes that the eyes, nose, and mouth are important in determining ethnicity in humans.

To determine the effectiveness of the results achieved, performance metrics such as sensitivity and specificity were calculated. Sensitivity

Table 6.3: Performance accuracy for the binary classification of face feature combination.

Feature Extraction Models	Linear Support Vector Machine (SVM)		
	Eyes & Nose	Nose & Mouth	Mouth & Eyes
ResNet-50	85.8%	*91.8%*	87.8%
ResNet-101	85.5%	*90.5%*	88.8%
ResNet-152	84.4%	*90.2%*	84.8%
VGG-16	81.5%	*85.2%*	82.8%
VGG-19	79.5%	*80.3%*	77.6%
VGG-Face	72.6%	*75.8%*	73.6%

is the value of positive cases classified correctly (TPR), and specificity is the value of negative cases correctly rejected (TNR). By combining sensitivity and specificity, the overall performance accuracy of the classification algorithm is reported. A visual representation of the values is shown in the form of a Receiver Operating Characteristics (ROC) Graph.

Figure 6.2 shows the ROC generated for the nose image experiment using the three tested deep learning algorithms; ResNet-50, ResNet-101, ResNet-152. While, Figure 6.3 shows the ROC for the VGG-based models, namely VGG-Face, VGG-16, and VGG-19 models.

Figure 6.2 shows the ROC for the nose images experiment using the three deep learning algorithms; ResNet-50, ResNet-101, and ResNet-152. All three models performed similarly (i.e., within the region of > 90%), which is depicted by the position of the performance lines that are positioned to the left of the axis. Despite having seemingly lower accuracy, RestNet101, having an area under the curve (AUC) of 0.9773 demonstrates competitive performance. This shows that even with limited facial detail, i.e., nose images only, the deep-layered architecture is sufficient at capturing ethnic traits.

Furthermore, Figure 6.3 shows the ROC for the isolated nose images using the VGG features. Unlike the ResNet models which are trained on

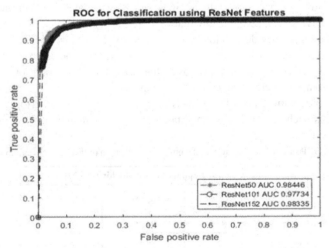

Figure 6.2: Receiver Operating Characteristics (ROC) Curve for the binary classification of the Pakistani nose, using Residual Neural Networks (ResNet-50, ResNet-101, and ResNet-152). The True Positive Rate (TPR) is the value of cases where the data has been correctly identified. False Positive Rate (FPR) is the value of cases where data has been incorrectly identified as positive.

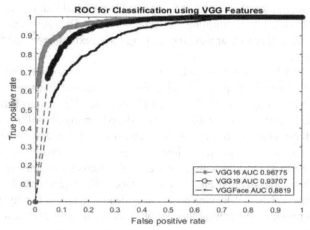

Figure 6.3: Receiver Operating Characteristics (ROC) Curve for the binary classification of the Pakistani nose, using VGG-based pre-trained models, namely VGG-16, VGG-19, and VGG-Face. The True Positive Rate (TPR) is the value of cases where the data has been correctly identified. False Positive Rate (FPR) is the value of cases where data has been incorrectly identified as positive.

non-face images, VGG-based models are trained specifically on a dataset of faces. It is apparent from the ROC curve that VGG-16 outperforms the other two models, and VGG-19 outclassed the VGG-Face. Such results are surprising, especially since they are trained on aspects of the face. This is especially the case when comparing both the ROC curves, deep neural networks extract relevant information, especially when there is very little facial detail, to yield a high-performance accuracy, in comparison to the VGG-based models.

As per Table 6.1, ResNet-50 achieved a performance accuracy of 94.1%, followed closely by ResNet-152 and ResNet-101 with classification at 93.6% and 93.4% respectively. The closeness in performance between the ResNet models is consistent with results from others. The 50/101/152-layer ResNets yield high and accurate results given the considerably increased depth of the architectures.

Due to the novelty of both our stimuli (i.e., isolated nose database) and the specificity of our approach (focused on ethnicity determination based on nose information only), it is difficult to directly compare the results here with the work of others. Nonetheless, the results suggest greater accuracy compared to those presented by Lv et al. [7] who utilised 3D nose shape classifiers. They report an ethnicity classification of 89.2% for the Asian and 87.4% for the White ethnic group.

Furthermore, the results here support the proposal that the nose is a useful feature for determining Pakistani ethnicity. This is in agreement with results other similar machine learning-based experiments conducted by others.

To the best of our knowledge, the nose as an isolated feature to address a binary ethnicity classification task has not been conducted previously, using either ResNet and VGG-features. The results achieved from all 8 experiments establish that the deep learning algorithms tested outperform VGG-based models. We suggest that this is due to the deep architecture of such models. Moreover, given the diversity and sexual dimorphism associated with the nose, in situations where humans may struggle to classify the ethnicity of a face and discriminate between Pakistani and non-Pakistani noses, deep learning methods may prove able to provide information that is not available from the human visual system.

Conclusions

In this chapter, we addressed the ethnic classification of physical facial features, especially the nose. A binary classification task was investigated using a novel dataset of nose images for the South Asian, Pakistani ethnic group. ResNet and VGG-features were extracted and fed to a Linear Support Vector Machine (SVMS) classifier. Performance accuracy of 94.1% was found for ResNet-50, whereas VGG-16 classified nose images with 90.8% accuracy. Our results are novel since the ethnic classification of the Pakistani nose has not been previously attempted.

Here, we have investigated the distinctiveness of the nose as a marker of ethnicity for the South Asian, Pakistani population. The results show that very deep learning algorithms outperform—specifically the VGG-based models, on the classification of isolated face feature images. The results presented here advance the current understanding of ethnicity determination because, to the best of our knowledge, the application of deep learning algorithms for the ethnic classification of Pakistani faces based on a single feature, specifically the nose, has not previously been attempted. Moreover, there is limited evidence available on the use of the nose biometric feature for the use of identification.

References

[1] Ozdemir, F., and A. Uzun. Anthropometric analysis of the nose in young Turkish men and women. Journal of Cranio-Maxillofacial Surgery 43(7): 1244–1247, 2015.

[2] Lu, X., and A. K. Jain. Ethnicity identification from face images. In Proceedings of SPIE - The International Society for Optical Engineering 5404, 2004.

[3] Hariharasudhan, D., S. Virasundarii, S. Selvamegala, S. Roomi, and S. Jeevanandham. Race Classification Based on Facial Features, in Computer Vision, Pattern Recognition, Image Processing and Graphics, pp. 54–57, 2011.

[4] Lewenberg, Y., Y. Bachrach, S. Shankar, and A. Criminisi. Predicting personal traits from facial images using convolutional neural networks augmented with facial landmark information. In Proceedings of the Thirtieth AAAI Conference on Artificial Intelligence, pp. 4365–4366, 2016.

[5] Qiu, X., Z. Sun, and T. Tan. Global texture analysis of iris images for ethnic classification. In: Zhang, D., and A. K. Jain (eds.). Advances in Biometrics. ICB 2006. Lecture Notes in Computer Science, vol. 3832. Springer, 2006.

[6] Chang, K. I., K. W. Bowyer, and P. J. Flynn. Multiple nose region matching for 3D face recognition under varying facial expression. IEEE Transactions on Pattern Analysis and Machine Intelligence 28(10): 1695–1700, 2006.

[7] Lv, C., Z. Wu, D. Zhang, X. Wang, and M. Zhou. 3D Nose shape net for human gender and ethnicity classification. Pattern Recognition Letters 126: 51–57, 2019.

[8] Drira, H., B. B. Amor, M. Daoudi, and A. Srivastava. Nasal region contribution in 3D face biometrics using shape analysis framework. pp. 357–366. In: Tistarelli, M. and M. S. Nixon (eds.). ICB 2009: LNCS 5558, 2009.

[9] Song, S., K. Ohnuma, Z. Liu, L. Mei, A. Kawada, and T. Monma. Novel biometrics based on nose pore recognition. Optical Engineering 48(5): 057204, 2009.

[10] Zehngut, N., F. Juefei-Xu, R. Bardia, D. K. Pal, C. Bhagavatula, and M. Savvides. Investigating the feasibility of image-based nose biometrics. 2015 IEEE International Conference on Image Processing (ICIP), 2015.

Further Reading

Bukar, A., H. Ugail, and D. Connah. Automatic age and gender classification using supervised appearance model. Journal of Electronic Imaging, 25(6): 061605, 2016.

Bukar, A., and H. Ugail. On automatic age estimation from facial profile view. IET Computer Vision 11(8): 650–655, 2017.

Bukar, A., and H. Ugail. A nonlinear appearance model for age progression. In: Hassanien, Aboul Ella, Oliva, and Diego Alberto (eds.). Advances in Soft Computing and Machine Learning in Image Processing (Springer). Springer, 2017.

Bukar, A., and H. Ugail. Facial age synthesis using sparse partial least squares (the case of Ben Needham). Journal of Forensic Sciences 62(5): 1205–1212, 2017.

Jilani, S., H. Ugail, A. Bukar, A. Logan, and T. Munshi. A Machine Learning Approach for Ethnic Classification: The British Pakistani Face, Cyberworlds, 2017.

Jilani, S. K., H. Ugail, S. Cole,. and A. Logan. Standardising the capture and processing of custody images. Current Journal of Applied Science and Technology 30(5): 1–13, 2018.

CHAPTER 7

Analysis of Skin Burns using Deep Learning

Surprising it may be, human skin burns are a common cause of death worldwide. Thousands suffer from burns and burn-related injuries every year. Records show that some 265,000 deaths are recorded globally every year due to burn-related injuries [1]. The leading causes of burn are flames, hot liquids, chemicals, electricity, and radiation from the sun. Thus, skin burns have become serious. Precise identification of burn injury is important to give the right medication.

Burn injuries can be excruciating depending on the extent to which the body tissue gets affected. Burns that affect top layers of the skin are less complicated than those affecting the deep tissues. Skin being the largest organ of the human body, an indirect consequence of a large burn on the body will directly affect the ability of the body to regulate heat, synthesis of vitamin D, and protect against infections.

Burn assessment is vitally important, as this will provide an insight into whether the burn will heal spontaneously or may require surgery. Visual inspection (known as clinical assessment) is the most commonly adopted means of identifying burns, but concern has been raised due to the poor accuracy of the technique. Moreover, the lack of medical professionals in most burn centres and the unavailability of aiding diagnostic tools have subjected thousands to danger. The study by [2], shows that wrong identification of burn severity leads to a poor assessment. The authors [2] stated that when a positive burn injury is falsely classified based on visual perception of the surgeon it may lead to unnecessary surgery, or prolong the patient's hospital stay. Similarly, the authors [2] have described the clinical assessment as an ineffective means to diagnose burn as it gives a maximum accuracy less or equal to 75%.

Researches have shown that a critical aspect in diagnosing burn relies heavily on experienced doctors and nurses. Besides, experienced burn clinicians have a challenge confronting them with burn patients every day, such as correct identification of burns that require skin grafting and those that do not. The problem here is likely misdiagnosing narrow burn injury and deep burn injury. The deep burn injury which affects flesh and damages muscles is the type requiring grafting in most cases. If the burn is underestimated and the right treatment is not offered, it may lead to irrecoverable complications, whereby allowing bacteria and other infections to contaminate the burn wounds. In the worst case, this could lead to the loss of life. Similarly, overestimating burn injury can lead to unnecessary treatment, requiring needless skin grafting and other irrelevant treatment.

According to [3], the accurate burn depth assessment is a challenging task even with experienced surgeons, but very important. Burn depth assessment with experienced surgeons was reported to be 60–75% accurate. Because of this, it was reported that Laser Doppler Imaging is the best technique to adopt for burn diagnosis over thermal imaging and vital dyes. However, the underpinning factor in adopting LDI is the cost and portability issues.

In another development towards improving medical care for burn patients, a study [4] was conducted for the burn depth assessment using LDI. Utilising LDI in this study was the fact that clinical means do not provide a satisfying result as expected. Forty patients were diagnosed with both clinical methods and LDI on different days after sustaining a burn injury. Assessment was carried out on 0,1,3,5, and 8 days using LDI where 54%, 79.5%, 95%, 97% and 100% accuracies were recorded. Clinical assessment provided 40.6%, 61.5%, 52.5%, 71.4% and 100% accuracies for the same number of days' assessments.

Authors in [5] were motivated and conducted a comparative study to determine the accuracies of LDI and clinical assessment in discriminating superficial and deep partial-thickness burn wounds. The study sample consisted of 34 patients of which 92 injuries were identified from March 2015 to November 2016. Features such as colour, dislodgement of hair follicles, fade, and pain were the parameters used to differentiate the different classes of burn depth. This was analysed statistically using the SPSS package. Depths of 57 wounds were correctly classified using a clinical approach,which provided an accuracy of 81.52% and a sensitivity of 81%. Similarly, 83 wounds were classified correctly using LDI with an accuracy of 90.21% and a sensitivity of 92.75%. Both methodologies have 82% specificity. LDI in burn depth assessment outperforms clinical assessment, but it takes a longer time and is very expensive.

Discrimination of Burnt and Healthy Skins

Here we discuss how deep learning approaches can be adopted for analysis of burn skin to discriminate between healthy skin and burns. We show how to use a fully trained convolutional neural network (CNN) to extract useful features which were subsequently fed into a Support Vector Machine (SVM) algorithm for classification. In this case, we show how features can be extracted from the 2D coloured images using the two variant deep CNN architectures, namely, ResNet101 and ResNet152.

The residual CNN by Microsoft Research Asia (MRA) proposed a residual network that led the contest at the ImageNet Large Scale Visual Recognition Competition (ILSVRC) 2015 classification exercise with an error rate of 3.57%. The unique feature of this network is the residual connection (shortcut connection). Thus, with the addition of input of the previous layer with an output of a lower layer, the resulting value then passes through an activation function which serves as the input of the next layer down the network. This provides the possibility of stacking networks with more layers with increasing accuracy. Skip connection was adopted after every few stacked layers, which is represented by a directed line labelled as a shortcut connection, as shown in Figure 1. Mathematically, we arrive at,

$$y = f(z) + (z) = f(z) + z. \tag{7.1}$$

The (z) is a shortcut that allows the gradient to be maintained and allows training the network quicker and with more layers stacked, and $f(z)$ is the output of the convolution layer, i.e., the layer from which the weights are derived. The residual network has two parts, namely the main part and the shortcut part. The main part comprises the regular convolution layer and the activation (ReLU) layers, while the shortcut part is the input directly connected to the output. Moreover, the activation function is applied after the output of the shortcut connection is added to the output of the main part. Thus,

$$y' = \sigma(f(z) + z) = \sigma(y), \tag{7.2}$$

where $\sigma(y)$ is the activation operation (nonlinearity) after the element-wise operation.

The general form of the residual network has 18,34,50,101 and comprises 152 layers. For the image analysis discussed here, this architecture and a 101 layer variant can be utilised. Since the model is pre-trained, one can resort to transfer learning to circumvent the computational complexity.

The SVM, as discussed earlier, is a learning algorithm that is primarily used for classification, regression, and outlier detection tasks on datasets. It works by finding an optimum separating hyperplane (OSH) which acts as a boundary that separates two classes of data. This is achieved by solving an optimisation problem. As a supervised learning algorithm, it learns to classify unseen data based on a set of labelled training data. Thus, the training phase is used to build a model, after that, new unseen data is mapped to one of the two learned classes by computing the dot product of the model (weights) and the image-data features.

Image Acquisition and Pre-Processing

Burn images were acquired which were directly captured from patients in the hospital, in addition to searches on the internet. A total of 600 human skin burn images and 200 healthy skin images (normal skin) were acquired. Out of the 600 images, 200 were identified and labelled as first stage burn images, 200 as second stage burn images, and the remaining portion as third stage burn images. All images were resized to 224 by 224 prior to the feature extraction and classification. This way, all images will be of the same size, which corresponds to the standard image size required by ResNet. Figure 7.1 shows an example image of a burn from the dataset.

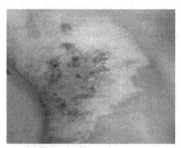

Figure 7.1: An example image of a burn from the dataset.

Evaluation of the Results

For the feature extraction from the images, the ResNet architectures, as discussed above, can be used. Furthermore, to test the validity of the results, a k-fold cross-validation where data is sub-sampled into k parts equally. Cross-validation (CV) is a statistical means of evaluating learning algorithms [9]. Training and validation are carried out by the subsamples. A special case of k-fold (where $k = 10$) cross-validation is used where $k - 1$ is used for training and the remaining split is used for the validation (testing). The iteration continues until all the subsamples are exhausted, in such a way that each subsample is used for both training and testing, and finally, the average of all the tests gives the result.

The result of the experiment is presented in this section. Table 7.1 shows the result obtained when the 101-layered architecture of the residual network used as a feature extractor and as an SVM classifier. In addition, Table 3 shows the result obtained using the 152-layered architecture of the residual network, and the same classifier was used as was in Table 7.2.

In Table 7.1, the predicted classes are the rows (predicted Classes), and the actual classes are represented by the column (Target Class). The correctly classified instances are represented by the leading diagonal, while the off-diagonal cells present misclassified observations. Similarly, the results from the ResNet-152 are presented in Table 7.2. Moreover, the performance of both the ResNet models is shown in Table 7.3.

Table 7.3 above shows accuracy and other performance metrics for the feature extractors. Both architectures achieved a total accuracy of more than 99%, but with 152-layered architecture, the accuracy is marginally higher with an overall achievement of 99.7% compared to 99.5% accuracy recorded by 101-layered architecture. This indicates that a network with deeper learning kernels can learn more features than a shallow network. This was compared to the findings in [6] as presented in Table 7.4.

Table 7.1: Classification results from the ResNet-101 layered architecture.

		Target Class	
		1	2
Predicted Classes	1	676	3
		49.7%	0.2%
	2	4	677
		0.3%	49.8%

Table 7.2: Result obtained using ResNet-152 layered architecture.

		Target Class	
		1	2
Predicted Classes	1	679	3
		49.9%	0.2%
	2	1	677
		0.1%	49.8%

Table 7.3: Performance comparison between ResNet-101 and ResNet-152 layered architectures.

	ResNet101	ResNet152
Precision	99.56%	99.56%
Sensitivity	99.41%	99.85%
Specificity	99.56%	99.56%
Accuracy	99.49%	99.71%

Table 7.4: Performance comparison of different CNN algorithms in discriminating between burn skin and normal skin.

	ResNet101	ResNet152	LeNet [55]
Precision	99.56%	99.56%	81.83%

Comparison of Popular CNN Models

Here we consider the implementation of other popular models for feature extraction and classification to compare their effectiveness for burn skin discrimination. We have chosen the VGG-16 and GoogLeNet, in addition to the results from ResNet models discussed earlier. The result displayed in Table 7.5 shows two classifiers used to classify the features extracted from the three different pre-trained models.

Firstly, the GoogLeNet pre-trained model was used for the feature extraction, and these features were fed into two different supervised classification algorithms (LSVM and DT). The classification result obtained from the classifiers shows that LSVM outperformed DT. The performance accuracies are approximately 99.3% and 92.3% for both LSVM and DT, respectively. Moreover, despite being trained and evaluated with the same datasets, the DT classifier is lagging behind LSVM in terms of accuracy with about a 7% error rate.

Similarly, classifying features extracted by the ResNet152 pre-trained model, LSVM recorded a classification accuracy of approximately 98.5% while DT achieved 93.4% as displayed in table 6. In this case,

Table 7.5: Performance accuracies of single base learners.

Feature Extractors	Classifiers	
	Linear SVM (LSVM)	Decision Tree (DT)
GoogLeNet	0.992647	0.922794
ResNet152	0.985294	0.933824
VGGNet	0.944853	0.926471

also a difference in the error rate produced by DT compared to LSVM is about 5.1%. The same intuition is applied to the scenario when the two classifiers were trained on features extracted by the VGG-16 model. LSVM achieved a performance accuracy of 94.5% while DT achieved a performance accuracy of 92.6% with a performance difference of 1.9% in favour of LSVM.

Improving the Learning and Classification

Here we discuss several ways we can improve learning and classification. In particular, we show how Ensemble learning, Bagging, and Adaboost can be utilised for this purpose.

Ensemble learning

Ensemble learning is a technique used to achieve a good result by using multiple learning algorithms in solving the same problem. In the ensemble learning paradigm, each single participating learning algorithm is referred to as a base learner, and it is believed that the aggregation of the generalisation of these learning is better than the generalisation obtained from a single base learner. This is since a single learning algorithm is capable of producing results better than the random guess, but when multiple base learners are used, the more accurate result is obtained, hence transforming the base learners (also referred to as weak learners) to strong learners. Ensemble learning which is also called committee-based based learning as stated by can be homogenous learning or heterogeneous learning.

- Homogeneous ensemble learning: Here, multiple weak learners or classifiers of the same type (same classification algorithms) are used
- Heterogeneous ensemble learning: A set of different classification algorithms are trained on the same training dataset to give an improved output compared to training a single classifier.

Ensemble learning techniques can be categorised into two: dependent and independent architectures. In the dependent architecture, the trained classifier is dependent on the output produced by its predecessor, which means the result obtained in the previous iteration guides the construction of the next iteration. But for the independent architecture, dependency on the prior output obtained from the previous iteration does not exist, and each classifier is trained on the dataset independently from each other.

Bagging

In this scenario, ensemble learner is constructed by training learning from a different subset of a given dataset, and the general prediction is obtained by voting across all predictions produced by the individual learners. The training data is generated by random sampling from the total data samples, say N. Bagging derives its name from **B**ootstrap **AGG**regating, where each random sample taken by each learner is referred to as bootstrap. However, the data selected is not completely removed from the dataset. This means data may likely appear multiple times in different training phases.

AdaBoost

This is a linear or sequential ensemble learning technique where weak learners are training one time. This implies that each base learner is trained at a time using the whole training dataset, and the next learner will be trained based on the residual error produced by the initially trained learner. Moreover, observations that were poorly classified are assigned more weight or given much priority when training the next model. This means that misclassified instances produced by the previous classifier are selected more often in the subsequent learning phase. Once the training is completed, the final accuracy is obtained by combining the results.

Evaluation of the Results

Having obtained the classification performances of the two employed classifiers in which both were trained on the features extracted by three different feature extractors, and the result shows that DT is the weaker classifier compared to its counterpart, then the idea of ensemble learning is applied using DT as presented in Table 7.6.

One hundred and fifty decision trees were assembled to boost or improve the accuracy of the poorly performed classifier in table 6. Bagging strategy with 150 DT was used in all three different cases as

Table 7.6: Performance accuracies of ensemble learners.

Feature Extractors	Classifiers	
	150 Trees (Bagging)	150 Trees (Boosting)
GoogLeNet	0.941176	0.977941
ResNet152	0.966912	0.970588
VGGNet	0.948529	0.963235

shown in Table 7.6. The DT obviously underperforms the obtained result when such a strategy is updated, as shown in Table 7.6. For each DT result from the three sets of features were extracted by GoogLeNet, ResNet152 and VGG-16, performance improvements of 1.8%, 3.3% and 2.3% were recovered using bagging.

Similarly, the boosting technique (AdaBoost) used has performed better than using a single DT classifier. Performance improvements are 5.5%, 3.7% and 3.7% using AdaBoost strategy trained on extracted features from GoogLeNet, ResNet152 and VGG-16. Boosting ensemble method has outperformed the bagging ensemble method in performance, as seen in Table 7.6. Note that, the bagging ensemble method operates by training multiple classifiers in parallel from the pool of datasets and voting is applied to determine the most occurred performance produced by the individual classifiers while boosting (AdaBoost) operates by training multiple classifiers one at a time, retraining next classifier on the output obtained by its predecessor, giving much more priority to the misclassified instances. This makes it much powerful than bagging, and the same argument was reported in different pieces of literature.

Moreover, it is worth noting that for each feature extraction model used, there is a likelihood of training and evaluating the model with a different set of data. The data are pooled from the same bank of images, but in each phase, the training and evaluation samples are drawn randomly. For instance, the burn image at the position i used during the training phase with GoogleNet, might not be used during the training phase with ResNet152. However, all four classification techniques were trained using the same features extracted by each model on the randomly selected data from the pool of images.

From the results obtained in both Table 7.6, the SVM algorithm has performed better in discriminating between healthy skin and burn skin than the decision tree and the ensemble techniques used. The interpretation is also well depicted, where single decision tree learners produced a poor result compared to ensemble learners. Additionally, the AdaBoost ensemble learning technique provides better performance improvement than the bagging ensemble learning technique. This is because AdaBoost operates by using a series of classifiers whereby each subsequent classifier is trained to remedy the misclassifications or errors made by the previous classifier.

Besides evaluating the effective accuracy of the classifier, some metrics need to be taken into consideration to complement the effective analysis of the result obtained. These metrics include how accurate is the classifier in predicting the actual burn images, i.e., those positive samples (sensitivity) and how accurate is the classifier in predicting healthy skin images, i.e., those negative samples (specificity). Sensitivity and specificity are reliable performance metrics in medical imaging which gives the complete picture. The result displayed in Table 8 is the specific performance metrics of all the classification carried out using GoogLeNet feature extractor. Similarly, Table 9 and Table 10 show similar performance metrics of the classifiers using ResNet152 and VGGNet feature extractors, respectively.

The results in Table 7.8, show that LSVM was able to accurately predict positive (burn) samples and achieved 100% sensitivity while slight misclassification in predicting negative (healthy) samples, 98.5% specificity was recorded. Similar to what was obtained in Table 7.6 and Table 7.7, ensemble learners outperformed single decision base learners in terms of correctly predicting positive samples (sensitivity) and negative samples (specificity) as displayed in Table 8 which was subsequently depicted in Figure 8. Both the single DT base learner and the 150-DT bagging ensemble learner-positive samples were more accurately predicted than negative samples wherein 150-DT boosting learning techniques, and the same performance accuracies were achieved for both sensitivity and specificity.

The LSVM, DT, 150-DT bagging learning technique and 150-DT boosting learning technique when trained on the features extracted by ResNet152, achieved sensitivities of 99.3%, 94.1%, 99.3%, and 97.8% each with corresponding specificities of 97.8%, 92.7%, 94.2% and 96.4%. The performance achievements of these classifiers were shown in Table 7.9.

Table 7.7: Area Under the Curve (AUC).

	GoogleNet	VGGNet	ResNet
LSVM	100.0%	99.1%	100%
DT	92.2%	92.9%	91.9%
150-DT (Bagging)	99.4%	98.9%	99.6%
150-DT (Boosting)	99.9%	99.3%	99.8%

Table 7.8: Performance metrics of the classifiers trained on features extracted by the GoogLeNet model.

METRICS	CLASSIFIERS			
	LSVM	**DT**	**150-DT (Bagging)**	**150-DT (Boosting)**
Sensitivity	100.0%	94.1%	96.3%	97.8%
Specificity	98.5%	90.5%	92.0%	97.8%

Table 7.9: Performance metrics of the classifiers trained on features extracted by the ResNet152 model.

METRICS	CLASSIFIERS			
	LSVM	DT	150-DT (Bagging)	150-DT (Boosting)
Sensitivity	99.3%	94.1%	99.3%	97.8%
Specificity	97.8%	92.7%	94.2%	96.4%

Table 7.10: Performance metrics of the classifiers trained on features extracted by the VGGNet model.

METRICS	CLASSIFIERS			
	LSVM	DT	150-DT (Bagging)	150-DT (Boosting)
Sensitivity	91.9%	94.1%	95.6%	96.3%
Specificity	97.1%	91.2%	94.2%	96.4%

Similarly, features extracted by the VGGNet model, as shown in Table 7.10 below, show a massive misclassification of a positive sample by LSVM where boosting ensemble learning technique recorded a better performance in predicting positive samples.

Analysis of Burn Depth

Recognition and discrimination of burns from healthy skin and other skin-related injuries is a first step in ensuring burns exist before going forward with the subsequent investigation. A huge volume of clinical burn images is accumulating every day as a result of image capturing devices in hospitals today and are processed by the human visual system. However, the interpretation of such images is prone to errors due to the fatiguing nature of humans.

Early identification of burn injuries and the precise evaluation of burn depth is of great importance to the patient towards the management of the injury, thereby ensuring early decision making on whether surgical intervention is required or not. Burn depth assessment is important because it can help the potential healing time of the identified burn wound, which allows proper treatment to be performed as early as possible. Burns that take a long period to heal such as superficial dermal and deep dermal burns can be treated as early as possible if they are identified which may minimise further complications due to assessment delay. Although identifying superficial epidermal and full-thickness burns is easy to identify clinically, can machine learning algorithms be used to differentiate superficial burns and full-thickness burns?

Skin is the first line of defence to the internal body organs, as such, it must be protected, and effective diagnosis should be ensured to tackle any possible threat. A burn can disrupt this protection functionality provided by the skin and allow the unnecessary flow of microorganisms into the body. Moreover, the injury inflicted on the skin depends on how severe it is. Does it affect the epidermis layer only, or does it extend to the inner layers such as dermis and hypodermis? Efforts have been made to provide an efficient means to detect skin diseases and injuries such as cancer. Therefore, we believe machines can also be used to provide first aid assessment of burns with high precision and accuracy because human judgement is believed to be subjective with a precision of 40–80%.

Clinical assessment is the most common method used to diagnose burns. Human judgement was found to be subjective and varies among surgeons with varying accuracies. The subjectivity of this method has been identified with two common problems—overestimation and underestimation. Overestimating burns may subject a patient to unnecessary skin grafting/surgery, whereas an increase in hospital delay and the possibility of subjecting the patient to a high risk of infection is associated with underestimating the burns. Histological assessment is fairly accurate than clinical judgement, however its limitation to only the biopsy site, and the probability of subjecting the patient to the risk of infection renders it unsuitable. Additionally, a biopsy is very painful. Additional and the most accepted method of burns assessment is the use of Laser Doppler imaging (LDI), which is objective rather than subjective nature of clinical means. It assesses the burn by measuring dermal perfusion. One of the advantages of adopting LDI as compared to histological assessment is its non-invasive feature. A study reported in [5] shows that LDI can assess burn healing time with up to approximately 90% precision. But it's a very expensive device, requires expertise to operate, and is also associated with healing times rather than the assessment of burn depth.

Based on supervised machine learning burn images can be classified into different categories. A dataset of 140 images captured using the digital camera was categorised into three classes of burns degrees (grade 1, grade 2, and grade 3). As a pre-processing step, the authors converted RGB images to grey-scale and used histogram equalisation to enhance the contrast of the images before cropping the images to size 90 by 90 pixels. k-Nearest Neighbour (k-NN) was used as the classifier. The results obtained were 65.5%, 82.5%, and 75% accuracies for correctly classifying grade 1, grade 2 and grade 3 skin burns. However, this study is associated with lots of diagnostic errors of up to about 25% [7].

Furthermore, authors in [8] lamented that there were reported cases in local medical centres where nurses had to make phone calls to a medical professional at a far location to describe the physical characteristics of the burn injury to be diagnosed. Similarly, classifying chemical burn images into either first, second or third degrees can be considered. Data were acquired using a camera of 10.0 Mp resolution, a total of just 120 images were acquired, where each category of burn has 40 images. The images were pre-processed first by cropping to size 90 by 90 pixels, features such as mean, discrete cosine transform (DCT), skewness, and kurtosis are extracted and used for the classification and obtained a slightly satisfactory result with an average accuracy of 88.33%.

Moreover, authors in [9] emphasised the need to provide a first aid assessment to burn patients, especially in local medical centres where there is a lack of well-trained medical professionals. They proposed a technique to identify and classify burns that heal spontaneously and those that need to be grafted using psychophysical experiments and multidimensional scaling analysis (MDS). They recorded 79.73% accuracy using 20 images. The perturbing part of this assessment using the human visual system is habituation error which is a tendency to generate the same result when presented with different samples, which may occur due to the subjective nature of humans.

Authors in [10] obtained burn images from different sources such as the internet, books, and direct capture in hospitals using digital cameras. These images were processed by transforming them to 90 by 90-pixel values, and colour features were extracted for training an algorithm for classification. The efficiency of 66%, 75%, and 90% was obtained using Template Matching (TM), k-Nearest Neighbour (k-NN) and Support Vector Machine (SVM) classifiers, respectively.

Diagnosing burns in dark-skin types was carried out by [11]. The study was proposed to find how effective experienced medical personnel could identify burns in dark-skin types. Data was collected using a Samsung Galaxy SIII mini from four different hospitals in Western Cape Town province, South Africa. These images were presented to physicians who were able to successfully diagnose size and depth of burn with 67.5% and 66.0% average accuracies respectively. However, the study focused only on burn identification rather than its categories.

In aggregate, most of the studies reviewed used traditional methods in extracting features or shallow networks. In the experiment section of this chapter, a deeper network that has a better computational efficiency with relatively more images o obtain an improved result in discriminating clinical burn images is used. The adoption of a deeper neural network is to ensure more relevant features are extracted which will enable it to have more distinguishable parameters. The use of more data images avoids overfitting problems, thereby making the algorithm generalise well for the unseen data.

Evaluation of the Results

The burn images were directly captured from patients in the hospital with full consent where a total of 34 images of different burn categories were obtained which were subsequently augmented by flipping, zooming, and rotation giving a total of 660 full-thickness burn images. Additionally, the superficial burn injuries (first-degree burns) were obtained via a search on the internet comprising mainly sunburn images. A total of 46 images of superficial burn images were successfully generated, then augmented to 660 images. Similarly, 660 healthy skin images were incorporated into the investigation. All images were resized to 224 by 224 prior to the feature extraction and classification. This way, all images will be of the same size which corresponds to the standard image size required by ResNet.

The experiment presented previously is inspired by the performance of Residual Network (ResNet) architecture as such the lower weights of the model were used (excluding the last layer) as filters which are convolved against 2D images to extract features. Afterwards, the features are fed into a linear support vector machine. In a nutshell, the approach is simple yet powerful, filters of fully trained state-of-the-art neural networks are used to convert raw image pixels into dense features, then, an SVM classifies the features into the non-burns superficial burn and full-thickness burns.

To ensure unbiased validation, k-fold cross-validation techniques are used where data is sub-sampled into k parts equally. Cross-validation (CV) is a statistical means of evaluating learning algorithms [9]. Training and validation are carried out by the subsamples. A special case of k-fold (where k = 10) cross-validation is used where k-1 is used for training and the remaining split is used for the validation (testing). The iteration continues until all the subsamples are exhausted, in such a way that each subsample is used for both training and testing and finally, the average of all the tests gives the result.

The contingency table, as shown in Table 7.11 (contingency table of confusion matrix) shows the summary of all the results obtained. The columns correspond to original classes fed into the algorithm, and the rows are the corresponding predicted outputs by the algorithm.

The labels (1 to 3), as shown in Table 7.11, mark the individual class category. Healthy skin images were labelled as 1, superficial burn images were labelled as 2, and full-thickness burn images were labelled as 3.

Table 7.11: Burn depth classification result.

		Target Classes		
		1	**2**	**3**
Predicted Classes	**1**	660 33.3%	0 0.0%	1 0.1%
	2	0 0.0%	660 33.3%	0 0.0%
	3	0 0.0%	0 0.0%	659 33.3%

True negatives (TN) are actual data of a class and are predicted as such. Similarly, True negatives are actual data classes that do not belong to a class and are predicted as such, for instance, normal skin and predicted as normal. When data in a class are negative but predicted as positive by the algorithm, it is referred to as false positive. But when data is positive and predicted by the algorithm as negative, then it is false negative.

The performance evaluation of the classification algorithm in categorising each instance class is presented in Table 7.12. The classifier mapped an instance from the three classes into a set {1, 2, 3} as healthy, superficial burn, full-thickness burn class labels. The performance metrics such as precision, sensitivity (recall), and specificity are computed as well.

Precision as a performance metric tells us how accurate a classifier is to recognise exact samples which were labelled. The recall is the number of accurately categorised positive samples divided by the number of positive samples contained in the data, whereas specificity is the measure of accurately identifying negative samples.

Moreover, the overall accuracy of the classification algorithm and the corresponding precision metric are shown in Table 7.13.

Table 7.12: Performance metric for healthy skin, superficial burn, and full-thickness burn

Metrics	Class 1	Class 2	Class 3
Precision (%)	0.99848	100.00	100.00
Sensitivity (%)	100.00	100.00	0.99848
Specificity(%)	0.99924	100.00	100.00

Table 7.13: Accuracy and precision for differentiating healthy skin, superficial burn, and full-thickness burn.

Overall accuracy (%)	Overall precision (%)
0.99949	0.99949

The results, however, recorded a misclassified element from full-thickness burns which is falsely predicted as superficial burn as contained in Table 7.13. Probably this misclassification could be due to the waxy appearances of some of the full-thickness images. Overall Correctness (accuracy) of the model is determined by summing the correct classifications (actual classified positives) divided by the sum of all classifications.

Conclusions

The investigation discussed in this chapter presents how a deep learning framework to automatically discriminate skin burn injury, and normal skin is possible. The obtained results show that machines can be used to give a first assessment in identifying human skin burn injury with high accuracy as compared to clinical assessment. This indicates that diagnosing burns can significantly minimise the waiting times in hospitals by improving the efficiency of service delivery. The conclusive objective of diagnosing burns is to provide an immediate assessment to reduce the length of hospital delays and to prevent loss of lives. Although efforts have been in place for the treatment of burn patients in clinical settings, the accuracy of the results obtained is not convincing, and above all, unreliable, because such methods are highly subjective.

Biopsy, despite being a better technique than clinical assessment, is subject to limitations such as being confined to a small region and ethical grievances due to leftover scarring. Similarly, Laser Doppler imaging proved to be a good technique for the treatment of burn wounds. However, in addition to the highlighted problems such as its weights and cost, studies have shown that LDI is only reliable and gives more accurate results from 48 hours after injury.

Therefore, in this research, we presented a concept of using machine learning to discriminate between healthy skin, superficial burns, and full-thickness burns. The result shows a classification accuracy of up to 99.9% using the 101-layered architecture of Residual Network (ResNet). This indicates that a more reliable and unbiased assessment can be achieved when such

Thus, the investigation in this chapter shows machine learning is capable of discriminating between burnt and healthy skin. Furthermore, the study has also shown that skin abnormalities that share similar physical characteristics which are very challenging to discriminate even with experienced dermatologists can be discriminated using computer algorithms effectively. Lastly, the study also shows two categories of burns injuries can be diagnosed effectively using a machine learning approach.

References

[1] Log, T. Modeling skin injury from hot rice porridge spills. International Journal of Environmental Research and Public Health 15(4): 808, 2018.

[2] Yang Zhao, Jason R. Maher, Jina Kim, Maria Angelica Selim, Howard Levinson, and Adam Wax. Evaluation of burn severity *in vivo* in a mouse model using spectroscopic optical coherence tomography. Biomedical Optics Express 6(9): 3339–3345, 2015.

[3] Monstrey, S., H. Hoeksema, J. Verbelen, A. Pirayesh, and P. Blondeel. Assessment of burn depth and burn wound healing potential Assessment of burn depth and burn wound healing potential. Burns 34(6): 761–769, 2008.

[4] Hoeksema, H., K. V. de Sijpe, T. Tondu, M. Hamdi, K. Van Landuyt, P. Blondeel, and S. Monstrey. Accuracy of early burn depth assessment by laser Doppler imaging on different days post burn. Burns 35(1): 36–45, 2009.

[5] Jan, S. N., F. A. Khan, M. M. Bashir, M. Nasir, H. H. Ansari, H. B. Shami, U. Nazir, A. Hanif, and M. Sohail. Comparison of Laser Doppler Imaging (LDI) and clinical assessment in differentiating between superficial and deep partial thickness burn wounds. Burns, 2017.

[6] Badea, M.-S., C. Vertan, C. Florea, L. Florea, and S. Bădoiu. Severe burns assessment by joint color-thermal imagery and ensemble methods. In e-Health Networking, Applications and Services (Healthcom), 2016 IEEE 18th International Conference, 2016.

[7] Sabeena, B., and R. K. Kumar. Diagnosis and detection of skin burn analysis segmentation in colour skin images. International Journal of Advanced Research in Computer and Communication Engineering 6(2), 2017.

[8] Abraham, J., M. Hennessey, and W. Minkowycz. A simple algebraic model to predict burn depth and injury. International Communications in Heat and Mass Transfer 38(9): 1169–1171, 2011.

[9] Serrano, C., R. Boloix-Tortosa, T. Gómez-Cía, and B. Acha. Features identification for automatic burn classification. Features identification for automatic burn classification. Burns 41(8): 1883–1890, 2015.

[10] Suvarna, M., and U. Niranjan. Classification methods of skin burn images. International Journal of Computer Science & Information Technology 5(1): 109, 2013.

[11] Boissin, C., L. Laflamme, L. Wallis, J. leming, and M. Hasselberg. Photograph-based diagnosis of burns in patients with dark-skin types: The importance of case and assessor characteristics. Burns 41(6): 1253–1260, 2015.

Further Reading

Abubakar, A., H. Ugail, and A. Bukar. Noninvasive assessment and classification of human skin burns using images of Caucasian and African patients. Journal of Electronic Imaging 29(4): 041002, 2019.

Abubakar, A., H. Ugail, and A. Bukar. Can machine learning be used to discriminate between burns and pressure ulcer? In Proceedings of SAI Intelligent Systems Conference, pp. 870–880, 2019.

Abubakar, A., H. Ugail, and A. Bukar. Assessment of Human Skin Burns: A deep transfer learning approach. Journal of Medical and Biological Engineering, 40: 321–333, 2020.

Abubakar, A., H. Ugail, A. Bukar, A. Aminu, and A. Musa. Transfer learning based histopathologic image classification for burns recognition. In 2019 15th International Conference on Electronics, Computer and Computation (ICECCO), 2020.

Smith, K., A. Abubakar, S. Jivan, D. Tobin, A. Mahajan, H. Ugail, and K. Poterlowicz. Can machine learning be used to extract relevant burn injury information automatically from 2D photographs? British Journal of Dermotology 80(6): E205–E205, 2019.

CHAPTER 8

Deep Learning Approaches to Cancer Diagnosis using Histopathological Images

In this chapter, we show an example of deep transfer learning. We demonstrate how deep transfer learning strategies can be utilised as a way of overcoming automated diagnosis challenges of tumours encountered in the digital pathology field. Through several experiments, the two strategies of deep transfer learning for image classification are investigated. The first strategy consists of using a network initialised with pre-trained weights and partially retraining it on a new dataset while the second strategy uses features extracted from pre-trained CNN without retraining the network and use them for training a third-party classifier. The experiments aimed to compare the performance of different CNNs based on deep transfer learning strategies for classifying Squamous Cell Carcinoma (SCC) tumours of the head and neck cancer from histopathological images. A total of 1,424 histopathological images were used to detect the tumours cells in sections. In terms of performance, both strategies have been compared thoroughly in histopathology images.

Here we compare the performance of different CNNs based on transfer learning with fine-tuning a network. We show experimental results from eight different CNNs—AlexNet, VGG16, VGG 19, GoogLeNet, Inception-V3, ResNet-50, ResNet-101, and InceptionResNet-V2. This is done by fine-tuning these pre-trained CNNs and by replacing the fully connected layer, which is configured for 1000 classes. The new layers for binary classes and then partially retraining them on the head and neck squamous cell carcinoma (HNSCC) tumour images.

Introduction

In pathology labs, tissues are sectioned, stained, and mounted on a slide of glass and then are put in the digital scanner for whole slide imaging (WSI) and visualising them at different magnifications. Due to the progress in medical scanning technologies, it would be possible to produce a significant number of digital glass slides which pass through a system for scanning and these slides can contain many different staining techniques with different kinds of tissues. Thus, the variability of tissue appearance and their quantity requires an efficient computer vision approach, which is the most difficult challenge in histopathology image analysis. The variability of tissue appearance is mostly the result of the variability in the conditions of the tissue preparation and staining processes. The other challenge is the lack of annotated data of digitised slides. Indeed, these annotations require an expert pathologist and are therefore difficult to obtain.

In parallel, Computer-Aided Diagnosis (CAD) methods based on deep learning techniques have recently had an impressive impact on the digital pathology field, which improved natural image recognition performances. The ongoing development in digital pathology allows for automated image analysis methods to support pathologists at such tasks and to increase the reliability of quantitative assessments.

Computer-Aided Diagnosis in Pathology

Over the years, there has been a marked change in pathology services with a significant shift from conventional pathology methods to digital pathology. These changes have been observed on both a national and international level. A rise in the use of digital technology has been observed in many areas of pathology within the UK. Specialist departments such as histopathology are increasingly utilising digital images along with other digital systems such as barcoding tools (used to mark and identify samples) to improve efficiency. Furthermore, these changes have also been observed on an international level due to the nature of digital technology which permits international collaboration; often required due to the geographical spread of patients and pathologists.

Several factors have contributed towards this prominent transition from conventional pathology to digital pathology; namely a rise in the pathology workload and a decrease in the pathology workforce. The Cancer Research UK reports that the ageing population has had a major impact on pathology workload. A significant link has been identified between the ageing population and cancer incidence rates. Since 2009 there has been a 16% increase (per annum) in the number of GP referrals for suspected cancer cases. In addition, in 2012 the World Health Organisation (WHO) reported an estimated 14 million new cancer cases with figures set to increase by a further 70% in the next two decades, thus, contributing towards an increase in pressure for pathology services.

Whilst the demand for pathology services is increasing, the pathology workforce is significantly decreasing. Once again, this can be observed on both a national and international level. The Cancer Research UK reports state that there will be a significant impact on pathology services over the next decade due to a decrease in the number of trainee pathologists. Similar findings have been observed in studies in the USA. For example, in the USA it suggested that there will be a significant decrease in the number of working pathologists over the next 20 years, with numbers set to decrease to 3.7 pathologists per 100,000 cases. An imbalance has been observed between the intake of new pathologists and the number of pathologists undergoing retirement. Retirement figures are set to soar by the year 2021, whilst intake numbers remain low.

Due to a decrease in the number of working pathologists, conventional pathology methods; which can often entail posting glass slides for consultation, can significantly impact turnaround times, i.e., impact on how long it takes for the patients to receive their lab results. As a result, the need for faster and efficient diagnostic methods such as computer-aided diagnosis has become a key area of interest.

CAD also referred to as computer-aided detection is the use of computer algorithms to assist the interpretation of medical images. Several steps are involved in computer algorithms, such as image processing, image feature analysis, and data classification. Features play a crucial part in image processing. A feature is defined as a single measurable characteristic of an image. To obtain features, the first step involves carrying out pre-processing procedures on images such as resizing. This is subsequently followed by feature extraction (image feature analysis) methods which are used to obtain features. The features are then used to recognise and classify images. Feature extraction is a large area of research and is a crucial step before image classification.

CAD has become increasingly popular over the years; playing a significant role in disease diagnosis. A rise in CAD technology has been particularly observed in radiology which is currently experiencing similar challenges to pathology. The demand for radiology services is significantly increasing whilst the number of radiologists is significantly decreasing. Double reviews are often required to ensure diagnostic accuracy and precision, however, due to limited numbers of radiologists and cost implications; this is not a viable option, thus explaining the rise in CAD technology.

Despite these positive results, several factors need to be considered before CAD systems can be fully integrated into clinical practice. This includes; the ability to provide fast and accurate results, improve the performance and efficiency of healthcare professionals, low implementation costs and save time. Currently, there are very few CAD systems that meet the requirements highlighted above, therefore are used as a means to provide a "second opinion" rather than a primary source for diagnosis.

Use of Deep Learning

Much work has been now begun in applying deep learning techniques to cancer diagnostics. Reproducibility is one advantage of this method. The ability to analyse entire slides in detail instead of focusing narrowly on specific areas of interest is also advantageous. Research has demonstrated that deep learning algorithms can achieve human-level pathologist levels of accuracy using CNNs, which are a class of deep learning concerned with architectures inspired by the structure and function of the brain. The Deep learning method for CAD has been reported to markedly outperform the classical machine learning method for biomedical image analysis. Compared to the classical machine learning method, the deep learning method, as stated in avoids subjective user bias because it does not require manual extraction of specific visual features of the tumour. Also, it does not require extensive pre-processing, segmentation, and feature extraction before classification. For example, taking patches of images instead of the feature vectors as input, one can perform classification based on the features of the image patch. The process learns the patterns of patch features from a large number of training patches. The outputs of CNN are the scores for different classes, and the class with the highest score is deemed as the classification result. Clearly, deep learning approaches have exceeded human performance in visual tasks by utilisation of automated hierarchical feature extraction and classification by multilayers, which could be applied for cancer diagnosis using tumour tissue slides.

While deep learning algorithms achieve state-of-the-art results in different machine learning applications, there are several challenges in their implementation in the biomedical domain. Firstly, training deep CNN requires a large amount of annotated images to learn millions of parameters, but these annotated images are currently lacking in the biomedical domain. Moreover, interpreting biomedical data requires an expert histopathologist. Therefore, it is expensive, time-consuming, and subject to observer variability. Secondly, training deep CNNs with a limited amount of training data leads to "overfitting", and features cannot generalise well on data. Overfitting is serious when the data contain high variability in the image appearance, which is usually the case in the biomedical domain. Finally, training deep CNNs from scratch requires high computational power, extensive memory resources, and time, so such approaches have practical limitations in the biomedical field.

Research has shown that CNNs can be used efficiently for deep transfer learning. In this case, a network can be trained using a source task and then be reused on a target task. This technique is very useful when the data for the target task is scarce. Therefore, deep transfer learning has been studied by the biomedical imaging community to overcome the data scarcity problem. Transfer learning techniques aim at reusing image representations learned from a source dataset with a large amount of labelled data on a second target dataset. Transfer learning techniques are shown to be a useful tool to overcome overfitting when the target dataset has a limited amount of labelled data.

One could, for example, define two transfer learning strategies for image classification, i.e., using fine-tuning pre-trained networks for a new task or off-the-shelf features extracted from pre-trained networks. The first strategy consists in using a network initialised with pre-trained weights and partially retraining it on the target task while the second strategy uses features learned from the source task without retraining the network and uses them for training a third-party classifier. Some of the first works of using the off-the-shelf feature extractor networks in the presence of limited learning samples have been done by Decaf and Overfeat. Similarly, the well-known AlexNet architecture has been used to extract the generic features which are then used by other methods such as the support vector machines (SVM) or random forests (RF) for final classification. More specifically, fine-tuning has also been investigated and compared with networks trained from scratch. The classifiers are trained on off-the-shelf features on a wide variety of histopathology image datasets. These include breast cancer, lymph node, cell nuclei, and tissue texture.

Here we show how the properties of transfer learning of CNNs can be exploited for head and neck classification in histopathological images. Three experiments were carried out over our histopathology image datasets. In the experiments, eight different CNN architectures were implemented for HNSCC classification.

Image Datasets

Dataset used in this report was obtained from The Ethical Tissue Department at the University of Bradford. The samples were collected from surgical operations performed in the period 2010–2012, from patients with head and neck tumours. These tumours were Squamous Cell Carcinomas (SCC). All slides were prepared according to the standard laboratory protocol that consists of formalin fixation and paraffin embedding of the Fresh Frozen (FF) tissue, followed by cutting off 3–5 μm thick sections, stained and mounted on a slide of glass.

All sections of the samples were stained with Hematoxylin and Eosin (H&E). The glass slides were scanned using the Philips Ultrafast Scanner at x40 magnification with a high-resolution image sensor (0.25 μm/pixel). Digital histological images were visualised using the Philips Image Management System. Figure 8.1 shows examples of binary classes, where the top row represents the normal histology images class, and the bottom row represents the HNSCC tumour images class.

A total of 1,424 histopathology images have been derived. These images were stored in the jpeg compression format. These images are with a 40X magnification level, and they did not magnify to other levels.

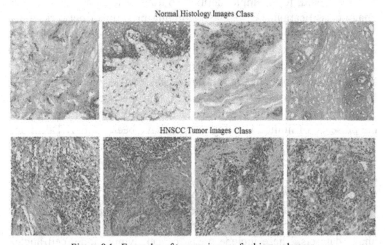

Figure 8.1: Examples of tumour images for binary classes.

Data Augmentation

One frequent problem in the field of machine learning is the lack of a sufficient amount of training data. One of the ways to deal with the problem is the so-called data augmentation. Image augmentation artificially creates training images. This is usually done through different ways of processing or a combination of multiple processing, such as random rotation, shifts, shear, and flipping, for example. It is a useful tool when training data is very little to boost the performance of deep networks.

Data augmentation is a fundamental step to obtain enough various samples and learn a deep network from the images. This is due to the difference in the number of images among different type classes and included random resizing, rotating, cropping, and flipping methods. The importance of image augmentation in deep learning is to get enough different samples that are needed to train CNNs from the images. Deep neural networks require a copious amount of training data to adjust their weights in a meaningful way and to achieve good performance.

Data augmentation is an essential step to have enough diverse samples and learn a deep network from the images. Small training datasets could cause the network to overfit. Often, the desired amount of data is much larger than what is feasible to acquire; this particularly applies to biomedical data. In such cases, data augmentation will be used to generate additional artificial data by modifying the existing dataset.

Method of Classification

This section explained the methodology of the 1st experiment. The last three layers of the pre-trained CNN are configured for 1000 classes. Therefore, these layers must be fine-tuned for the new binary classification problem (HNSCC tumour and normal histology). To achieve that, all the layers were extracted, except the last three, from the pre-trained CNN and then retrained them to classify head and neck tumour. Therefore, the weights of CNN were preserved while the last three layers were updated continuously. Figure 8.2 is an example of how to apply this methodology for the VGG-16 model.

As explained earlier, we took a total number of 1,424 images were fed into the CNN model, by resizing the images according to the size of the image input layer. The AlexNet model requires input images of size $227 \times 227 \times 3$, where 3 is the number of colour channels. VGG, GoogLeNet, and ResNet architectures require input images of size $224 \times 224 \times 3$, while Incv3 and IncResNet models require input images of size $229 \times 229 \times 3$. After that, the images were divided randomly into a training set of 80% (1,140 images) and a testing set of 20% (285 images), and then they were rotated (90, 180 and 270), flipped left to right horizontally, and then vertically. The small overlapping patches of sizes 32×32 were cropped along the images. Data augmentation usually helps to prevent the network from overfitting and to memorise the exact details of the training images.

For transfer learning, all layers from source models were copied to our target networks except the last fully connected layer. Then the last fully connected layers were modified for adapting the models to

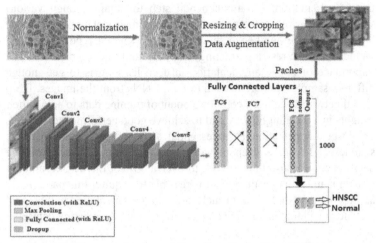

Figure 8.2: Diagrammatic representation of how fine-tuning on VGG-16 can be carried out.

our binary HNSCC classification task. To learn faster in the new layers compared to the transferred layers, one can add an increased Learn Rate Factors for the fully connected layer and set the initial learning rate to .0001.

For the network training, all network models were trained for 15 epochs with a batch size of 128. The training was performed using Stochastic Gradient Descent with Momentum set to 0.90. Here, the learning rate was initially set to 0.0001 as a starting point and was decremented after each update. The program validated the network every three iterations, and 120 iterations were selected as the maximum number. For a fair comparison, the different approaches were trained and tested with the same training sets and testing sets. Statistical performance measurement results from implemented different deep learning models were summarised in Table 8.1.

Off the Shelf Feature Extraction

The off the shelf feature extraction strategy is a way to use pre-trained networks to extract learned features and then used them to train a third-party classifier without consuming time and effort into training. Here, four common machine learning algorithms for performing classification included Support Vector Machines (SVM), k-Nearest Neighbor (kNN), Random Forests (RF), or a fully-connected single layer perceptron (FC). Deep learning models were used to build these machine learning classifiers.

Table 8.1: Standard Performances from fine-tuning the last layer for different models.

Deep Learning Model		Standard Performance Measurements			
		Accuracy (%)	Precision (%)	Sensitivity (%)	Elapse Time (min)
Series Network	AlexNet	98.25	98.74	99.16	15
	VGG16	95.44	97.86	96.63	25
	VGG19	95.79	97.47	97.47	30
DAG Network	GoogLeNet	95.09	96.65	97.47	45
	IncV2	93.33	93.25	99.16	120
	ResNet50	97.89	98.73	98.73	90
	ResNet101	96.84	98.72	97.47	200
	IncResNetV2	88.42	87.78	1.00	300

Here we explain how the discriminant features can be extracted from the activation layer of the fully connected layer of CNN and then were used to train SVM classifiers, such as SVM, kNN, and RF for classification problems (HNSCC tumour and normal histology). Discriminant features can also be extracted from the activation layer of different layers of CNN and then were used to train the SVM classifiers. Figure 8.3 describes how to carry out off-the-shelf feature extraction. The features were extracted from the last connected layer (FC8) of the VGG-16 model. Images datasets were fed to the classifier to train it.

Experiment 1

SVM classifiers are an extremely popular and well-researched class of supervised learning models, which can be used in linear and non-linear classification tasks. In this implementation, the kernel function is used for mapping data points to feature space. The kernel function plays an important role in the classification process, so this experiment examines if the Linear or Gaussian function—also known as a Radial Basis Function (RBF) function—is more suitable for our application. Table 8.2 shows the results obtained from the experiment.

Experiment 2

The AlexNet, VGG16, and VGG19 architectures contain three fully connected layers namely, FC6 and FC7 with 4096 nodes (features), and FC8 before a final Softmax classifier, with 1000 nodes (features). Note that Softmax normalises fc8 to the probabilistic label vector of the CNN. Before that normalisation, the FC8 can be interpreted as an

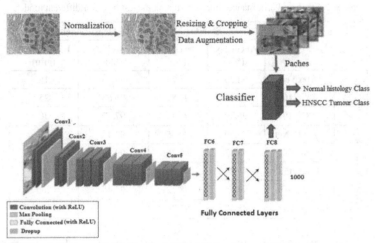

Figure 8.3: The technique that can be adopted for off-the-shelf feature extraction.

important feature vector for classification. Since each layer represents a different set of abstract features, it is important to choose one that best contributes to a well-performing classifier. Therefore, in this experiment, we examine which layer is more suitable for our application. Table 8.3 shows the results obtained from the experiment.

Experiment 3

The eight pre-trained CNN networks were used to feed four classifiers, and the performance results were reported in Table 8.4. For each classifier, the extracted features from the fully connected layer of the different eight pre-trained CNNs were used to train it. The classifiers used a linear function as the kernel function, and the sequential minimal optimisation method to find the separating hyperplane. We set its average kernel size to 2.0. The RF classifier was trained with 50 trees on the generated feature vectors from each approach. All the other parameters were not optimised and taken from the standards of the RF implementation in R.

Discussions

In this chapter, we have shown how it is possible to empirically investigate various deep transfer learning strategies for the analysis of histopathological images. The experiments discussed here showed that the AlexNet and ResNet architectures yield the best performing models across the fine-tuning network strategy. It appeared that SVM classifiers are the best performing feature extractor strategy followed by the RF. It

Table 8.2: Standard Performance Measurements for SVM Classifiers.

	CNN Models	Feature Layer	Accuracy (%)	
			SVM Linear Classifier	SVM Gaussian Classifier
Series Network	AlexNet	FC6	98.51	93.58
		FC7	98.68	93.75
		FC8	94.15	89.44
	VGG16	FC6	94.59	87.97
		FC7	94.05	87.47
		FC8	93.45	86.91
	VGG19	FC6	95.53	87.89
		FC7	92.56	85.16
		FC8	90.59	83.34
DAG Network	GoogLeNet	'loss3-classifier'	94.55	90.76
	Incv3	'predictions'	94.95	90.49
	Res50	'FC1000'	96.03	93.15
	Res101	'FC1000'	96.52	93.33
	IncResNet	'predictions'	98.00	94.75

Table 8.3: Performance Measurements for Classifiers using FC6, FC7, and FC8 Layers.

	CNN Models	Feature Layer	Accuracy (%)			
			SVM Linear Classifier	kNN Classifier	RF Classifier	FC Classifier
Series Network	AlexNet	FC6	98.51	92.27	94.54	83.16
		FC7	98.68	96.02	94.65	83.22
		FC8	94.15	90.59	91.68	75.24
	VGG16	FC6	94.59	90.58	93.06	86.62
		FC7	94.05	88.60	92.70	66.82
		FC8	93.45	85.14	92.07	83.16
	VGG19	FC6	95.53	91.57	96.52	75.94
		FC7	92.56	91.08	94.18	75.73
		FC8	90.59	89.90	93.55	73.26

also shows that the last layer features are always outperformed by features taken from an inner layer of the network. Furthermore, in general, the achieved accuracy of SVM, kNN, and RF classifiers extracted features from AlexNet and ResNet101 models were high. Therefore, among three

Table 8.4: Performance Measurements for Classifiers using different CNN models.

	CNN Models	Feature Layer	Accuracy (%)			
			SVM Linear Classifier	kNN Classifier	RF Classifier	FC Classifier
Series Network	AlexNet	FC7	98.68	96.02	94.65	83.22
	VGG16	FC6	94.59	90.58	93.06	86.62
	VGG19	FC6	95.53	91.57	96.52	75.94
DAG Network	GoogLeNet	'loss3-classifier'	94.55	80.74	90.10	64.85
	Incv3	'predictions'	94.95	94.05	97.02	88.60
	Res50	'FC1000'	96.03	91.57	94.06	87.61
	Res101	'FC1000'	96.52	96.03	98.01	82.17
	IncResNet	'predictions'	98.00	92.58	98.50	70.29

classifiers AlexNet and ResNet101 models were suitable for our image dataset.

Although the size of the image dataset was small, the results were shown to have good accuracy among different models. A highly successful classification has been achieved by the AlexNet model with an accuracy rate of %98.25, followed by ResNet50 and ResNet101 models with accuracy rates of %97.89 and %96.84, respectively. But VGG16, VGG19, GoogLeNet, and IncV3 performed classification with close accuracy rates of %95.44, %95.79, %95.09, and %93.33, respectively, and the IncResNet model was the worst one.

Comparing the timing performance of the eight methods, it is noteworthy that the series networks were faster than DAG networks. AlexNet was the fastest, it took just 15 minutes to train, followed by VGG16, VGG19, GoogLeNet, and ResNet50. However, IncV2, ResNet101, and IncResNetV3 took a long time, using a computer with a 2.2 GHz dual-core i7 CPU.

The results indicate that there was no difference between the precision of AlexNet, ResNet50, and ResNet101. They had the close and the higher correctly classified rates than others were 98.74 %, 98.73, and 98.72%, respectively. Although the accuracy and precision rates of the IncResNet model were the worst compared to other models, it had the highest sensitivity (recall) rate.

Thus, we show how popular such as AlexNet, VGG16, VGG19, GoogLeNet, IncV3, ResNet50, ResNet101, and IncResNetV2 models favour when using the fine-tuning strategy on histopathology image datasets. AlexNet and ResNet models were found more accurate, precision and reliable than other models. They had high sensitivity ranged from 99.00% to 97.00%.

Concerning evaluation, the classifiers in the three experiments were evaluated according to standard performance criteria of accuracy. To evaluate the various approaches, the same images for training and testing were used. The comparison was conducted using two times 10-fold cross-validation. For each cross-validation, the performance values were calculated for each feature set based on the nine folds of training samples, via grid search in the parameter space. Therefore, each cross-validation might have slightly different values, and the average optimal value was reported.

It is important to highlight that there is no consensus on whether the fine-tuning strategy is better than the off-the-shelf features extraction strategy. In terms of performance, both strategies have not been compared thoroughly yet in biomedical imaging, and there is no consensus about whether one is better than the other. Some studies in the biomedical field have shown that fine-tuning outperforms off-the-shelf features. However, this cannot be completely conclusive. Moreover, fine-tuning, as stated, is more computationally demanding than using the off-the-shelf features and involves dealing with more hyper-parameters. Whereas the off-the-shelf features approach extracts the features from CNN is usually fast and requires only tuning the hyper-parameters of the final classifier, which can usually be done efficiently through cross-validation.

Another key observation made in this study when considering the VGG-16 results is the number of incorrect predictions made at particular FC layers. Out of the 57 incorrect predictions made by the model using SVM of these incorrect predictions were made using FC layer 36 and 17 incorrect predictions were made using FC layer 34. The model only generated a 100% performance accuracy rate using FC layer 35 along with SVM and both FC layer 35 and SVM provided slightly better results compared to the other layers. Although it cannot be confirmed whether there is a potential link between using a particular FC, it is an area worth exploring. Furthermore, the different layers used in this project are also an area of interest. Layers in a model relate to particular characteristics of an image and are used to convert images into a set of meaningful numbers. However, which layer is related to what characteristic is still under investigation. More knowledge on the layers could help answer the question as to why certain layers (FC layer 35) performed better than others (FC layer 34).

The results from the different FC layers can also be used to assess in identifying tumour and non-tumour images. Information was extracted from the VGG-Face model and fed into a new algorithm. To ensure that the software would not generate a result from the VGG-Face model, FC layer 36 was used as the last FC layer. If an FC layer performed better close to the last layer of the VGG-Face model (FC layer 37), then this would suggest that the software performs better when image features are extracted very close to the last layer (which is more specific to faces). However, if FC layers further away from the last layer performed better, then it can be assumed that these layers learn more generic information and therefore facial recognition algorithms can be successfully used on any other images, such as tumours. As FC layer 36 was used as the last layer for the software, all 12 results from the study confirm that facial recognition algorithms can be used to successfully identify tumour and non-tumour images. However, it should be noted that FC layers 33 and 35 performed better than FC layer 36, thus suggesting that using lower FC layers in facial recognition algorithms generates better results.

Deep learning methods have proven to be extremely effective in learning for a wide variety of purposes. They are different compared to most machine learning models. However, as most models emphasise learning a complex classifier, whereas CNNs emphasise on learning complex features. Indeed, all of the layers of the network help to extract and learn increasingly complicated features. This is except for the output layer, which serves as a relatively simple classifier based on these features.

On the other hand, CNNs usually need large datasets for training that is why the recent approaches trend to transfer learning strategies that use the existing models which pre-trained on large image datasets. CNNs can be used efficiently for transfer learning, i.e., a network can be trained on a source task such as an ImageNet dataset and then be reused on a target task such as a histopathology image dataset. These trained models can then further be fine-tuned to a new dataset for better fitting or extracted features from their activation layers to feed the classifier.

Transfer learning technique, as stated in, is useful, especially when the data for the target task is scarce. For this reason, deep transfer learning has been studied by the histopathology field to overcome the data scarcity problem. There are essentially two transfer learning strategies, i.e., using off-the-shelf features extracted from pre-trained networks and fine-tuning pre-trained networks for the task at hand. The generalisability property of CNN makes their features transferable to other applications which encouraged the researchers to employ transfer learning for histology images. As can be explained, these features have also been used to train separate classifiers for predictions, which are particularly useful when there is not enough dataset for training the CNN from scratch.

Further Reading

Donahue, J., Y. Jia, O. Vinyals, J. Hoffman, N. Zhang, E. Tzeng, and T. Darrell. Decaf: A deep convolutional activation feature for generic visual recognition. In International Conference on Machine Learning, pp. 647–655, 2014.

Tajbakhsh, N., J. Y. Shin, S. R. Gurudu, R. T. Hurst, C. B. Kendall, M. B. Gotway and J. Liang. Convolutional neural networks for medical image analysis: Full training or fine tuning? IEEE Transactions on Medical Imaging 35(5): 1299–1312, 2016.

Ugail, H., M. Alzorgani, and S. Betmouni. A Deep Learning Approach to Tumour Identification in Fresh Frozen Tissues, 13th International Conference on Software, Knowledge, Information Management and Applications (SKIMA), 2019.

CHAPTER 9

A Deep Transfer Learning Model for the Analysis of Electrocardiograms

A cardiac attack or myocardial infarction (MI) is one of the common heart disorders. It occurs when one more coronary artery of the heart becomes blocked. Early detection of MI is essential for the reduction of the death rate that is rising. Currently, cardiologists use an electrocardiogram (ECG) as a tool that is a diagnostic monitor and reveals the MI signals. However, all the MI signals are not stable and can be noisy, and as a result, it can be challenging to detect or observe these signals manually. Moreover, manual analysis of large amounts of ECG signal data can be tedious and time-consuming. Thus, there is a need to develop efficient methods to automatically analyse the ECG data to arrive at an accurate diagnosis. In this chapter, we show how deep learning-based solutions can help to address this challenge. We show how two types of transfer learning techniques can be to retrain the pre-trained VGG-16 model to obtain two new networks, namely, VGG-MI1 and VGG-MI2. Specifically, we modify the last layer of the VGG-16 and the final layer of the VGG-Net model accordingly to suit our requirements. In addition, we show how various functions can be introduced to the model to reduce overfitting. Within the VGG-MI2, one layer of the model is selected as a feature descriptor of the ECG images to describe some of the essential features. Thus, in this chapter, we show how it is possible to develop an accurate tool for the analysis of electrocardiograms for the efficient diagnosis of MI.

Introduction

Coronary heart diseases are one of the primary causes of human deaths in the world. Over 17.7 million people die every year due to cardiovascular-related diseases. Heart attacks amount to over 70% of such deaths [1]. The leading cause of heart attacks is the inadequate ability for blood to flow within the heart's coronary arteries, causing a medical condition known as myocardial infarction (MI).

The functionality of the heart highly depends on the circulation of the oxygenated blood within it nourishing the arteries and the cardiac muscles. MI occurs when a coronary artery is so severely blocked, leading to an insufficient supply of nutrients and blood that is oxygen-rich. Commonly, sudden death could occur within an hour of the beginning of detecting the activities of MI [2]. Consequently, it becomes crucial to monitor the heart rhythms of suspicious subjects regularly to understand the condition of the heart and prevent any occurrence of MI.

The analysis of electrocardiogram (ECG) is one of the ways of providing adequate information to diagnose MI. Other standard tests are diagnostic magnetic resonance imaging (MRI) and echocardiography. Analysis of the ECG, however, is the most efficient and cost-effective way to provide an accurate diagnosis of MI. Moreover, it is often done through manual means whereby a suitably qualified clinician has to undertake a visual inspection for delivering a diagnosis. However, this process can be cumbersome, time-consuming, and potentially prone to errors. Therefore, the use of an efficient computer-based method to assist with this problem appears to be plausible.

In the past researchers have tried to address this problem using visual computing-based techniques. For example, Sadhukhan et al. [3] attempted to analyse the Harmonic phase distribution of the ECG signals, through image processing tools, to quantify the level of MI with an accuracy value of 95.6% [3]. Jayachandran et al. utilised wavelets and their multi-resolution properties for MI detection from ECG signals with an accuracy of 96.1% [4]. Dohare et al. utilised support vector machine (SVM) classifiers for analysing ECG signals to achieve 96.66% accuracy in the diagnosis of MI. Similarly, Arif et al. implemented k-nearest neighbour (kNN) classifiers for extracting time-domain features of the ECG signals to detect and localise the MI signals [5] achieving 99.9% accuracy in MI detection.

Machine learning being a popular methodology for image and signal analysis, has found its way in attempting to analyse ECG signals from the heat that have been reported in the academic literature recently. However, much of the methods are based on small-scale computationally intense machine learning models and feature extraction/classification. As a result, model overfitting and computational inefficiencies are often reported. The scope for deep learning to be utilised in this area appears to be promising. Hence, the aim here is to explore and showcase this possibility.

Several approaches have also been proposed to detect MI by digital analysis of ECG signals [6]. However, few of the previous works based on CNN have been utilised for MI detection. For example, Wu et al. [7] utilised a soft-max regression as a multi-class classifier for ECG signal analysis. To this framework, they then incorporated discrete wavelet transforms, which enabled them to learn features in the signals. With this formulation, they have demonstrated that it was possible to classify ECG signals for MI with an accuracy of 99.82%. Similarly, Acharya et al. [8] utilised a handcrafted 11 layer, single-dimensional CNN with noise filters to achieve an accuracy of 93.53% for MI detection. Liu et al. [9] proposed a method called multiple-feature-branch CNN (MFB-CNN) for MI detection with an accuracy of 98.79%. However, in all this work, the researchers have utilised large convolution filters leading to a heavy toll on the computational cost. Much of this work also suggest that their experiments were conducted on small-scale ECG datasets whereby the diagnosis can often result in low values of specificity and sensitivity.

A Transfer Learning Model

Here, we discuss a deep transfer learning architecture that can be successfully utilised for analysing ECG signals for the automatic diagnosis of myocardial infarction. The model is based on the VGG-16 base architecture to which several enhancements were made. The model discussed here has two forms, namely VGG-MI1, which is a new network structure, and VGG-MI2, which can be utilised for feature extraction from the ECG signal images.

Thus, using the pre-trained VGG-16, we obtain two new networks— VGG-MI1 and VGG-MI2—with a small filter size in the first convolutions layers and also a small number of pooling layers. This allows the new model to be more computationally efficient. Furthermore, the 1000 unit Softmax within VGG-16 is replaced by more straightforward 2 unit Softmax layers. Thus, the output of the new model is binary, which is suitable for the analysis of the ECG signals dealt here.

Table 9.1 provides details of the various layers in the proposed CNN model. Here we have, convolutional layers, pooling layers followed by 2 fully connected layers to the signals in image format ($128 \times 128 \times 1$) are fed into. The max-pooling layer enables us to control the complexity of the computation and helps with the problem of overfitting. In addition to this, non-linearity is also introduced to the model by providing the layers with an appropriate ReLU activation function such that,

$$f(x) = \max(0, x), \qquad (9.1)$$

where x is considered to be the input to the neuron.

To reduce both the noise and the computational cost, one could also employ 3×3 filters in the first convolutional layers as opposed to the commonly used larger filters (e.g., 5×5 or 10×10). Additionally, to avoid overfitting during the training phase, a dropout technique [10] is adopted in the fully connected layers of the network. The dropout was not applied to convolutional layers because these layers have a relatively small number of parameters compared to those in the activation layers.

To further optimise the proposed CNN architecture, here we used two subnetworks with enhanced transfer learning. We replace the last 1000 unit Softmax layer of the VGG model with a 2-unit Softmax layer, which we refer to as VGG-MI1. Details of the VGG-MI1 are shown in Figure 9.1. We then select the output of the second fully connected layer of the pre-trained which contains the feature descriptors. This network is referred to as VGG-MI2, as shown in Figure 9.2. Besides replacing the last 1000 unit soft-max layer, we also modified the various functions of the standard VGG model in the design of the VGG-MI1. This helps overcome the overfitting problem and improves classification accuracy.

Table 9.1: The architecture of the proposed VGG-MI.

Layers No.	Type	Kernel size	No. Kernels	Stride	Input size
Layer 1	Conv1	3×3	64	1	$128 \times 128 \times 1$
Layer 2	Conv2	3×3	64	1	$128 \times 128 \times 64$
Layer 3	Pool	2×2		2	$128 \times 128 \times 64$
Layer 4	Conv3	3×3	128	1	$64 \times 64 \times 64$
Layer 5	Conv4	3×3	128	1	$64 \times 64 \times 128$
Layer 6	Pool	2×2		2	$64 \times 64 \times 128$
Layer 7	Conv5	3×3	256	1	$32 \times 32 \times 128$
Layer 8	Conv6	3×3	256	1	$32 \times 32 \times 256$
Layer 9	Pool	2×2		2	$32 \times 32 \times 256$
Layer 10	Full		2048		$16 \times 16 \times 256$
Layer 11	Full		2048-dropout		2048
Layer 12	Soft		2		2048

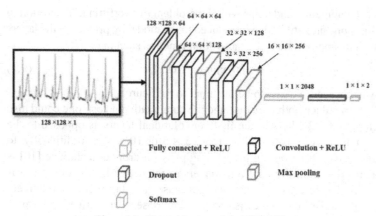

Figure 9.1: The architecture of the VGG-MI1.

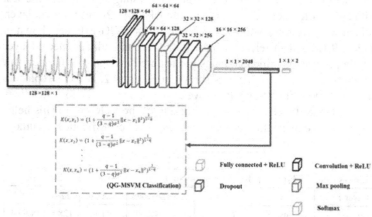

Figure 9.2: The architecture of the proposed VGG-MI2 model.

Training

To train the VGG-MI1 model, one could utilise the method of stochastic gradient descent (SGD) with a mini-batch. The size of these mini-batches can be 5, which can be enhanced by a momentum function. Additionally, the weights of the network filters can be initialised using random sampling. Here, for the random sampling, a Gaussian distribution with 0.01 standard deviation and zero mean was used. Furthermore, the hyper-parameters are set such that the learning rate is taken to be 0.001. The weight of the decay parameter is taken to be 0.0005. Table 9.2 shows the various parameters that can be utilised with the proposed CNN model. With these settings, the training and testing of the CNN can be taken

Table 9.2: The most important VGG parameters in our model.

Parameter	Value
Learning rate	0.001
Standard deviation	0.01
Mean	0
Weight decay	0.0005
Epochs	50
Minibatch size	5 samples

forward with a given number of epochs–usually, 50 epochs are sufficient to obtain an accurate network.

To train the VGG-MI2 model, one could use the Q-Gaussian multi-class support vector machine (QG-MSVM) classifier [11], which is defined as,

$$K(x, x_i) = \left(1 + \frac{q-1}{(3-q)\sigma^2} \| x - x_i \|^2 \right)^{\frac{1}{1-q}}, \qquad (9.2)$$

where q is a real-valued parameter. Furthermore, σ is a real value standard variance of Gaussian distribution, and each $x_i \in R_p$ is a p-dimensional real vector. In a different setting, the QG-MSVM was utilised to classify fingerprints, and it has shown to be effective when compared to the SVM kernels. Thus, we modified the QG-MSVM can be used with the settings such that, $\frac{1}{\sigma^2}$ is assigned to be 0.5 and q to be 1.5.

Dataset Utilised

To test the accuracy of the proposed CNN model, data from the PTB dataset was run on the models. This dataset is widely used in MI research. The PTB dataset has 549 EEG recordings of 290 subjects of which 28% are females, and 72% are males. The age of the subjects ranged from 17 years to 87 years. Out of the 549 subjects and 368 had clinically confirmed MI. For the experiments discussed here, a two-second duration of Lead II ECG signals was utilised. There is a total of 21,092 normal ECG beats along with 80,364 MI ECG beats were used. Here, each of the two-beat signals is represented as a 128 × 128 greyscale image. Figure 9.3 shows sample images of the signals for two seconds for normal and with confirmed MI.

Thus, the original EEG signals are transformed into two-dimensional ECG images by plotting each ECG signal as an individual 128 × 128

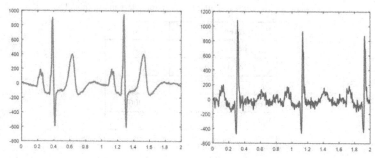

Figure 9.3: Sample Normal and MI ECG signals.

greyscale image. This is to ensure that the convolutional and pooling layers in the model are fully utilised to obtain better computational efficiency and accuracy. Additionally, working with images helps with data augmentation—a step often required to enhance the dataset for training deep learning networks.

For data augmentation, we divided each of the images into 9 smaller images so that each of the nine images represents 75% of the original image. Thus, we extracted four images from each of the corners, one from the left, one from the right, one from the bottom centre, and one from the top centre. Following that, the augmented images were resized so that they are also of size 128 × 128. Hence, with the proposed augmentation step, we were able to extend the dataset by 10 fold.

Measuring the Performance

To verify our model, we performed a test on the CNN model after every completed round of training epoch. The data were separated into three parts—60% of the data for training, 30% for validation, and 10% for testing. Moreover, we employed a ten-fold cross-validation approach, i.e., the performance of the system is evaluated at each fold, and the average of all ten folds was calculated as the final performance of the system.

The result from the CNN model can be presented in four ways. For the best-case scenario, we can look at the MI signals without any noise, and we can employ the augmentation technique on the trained data. For a second scenario, we can work on MI signals without noise and without using the augmentation technique. For the third scenario (i.e., the worst case), we can work on the MI signals with noise and using the augmentation technique on the trained data. Finally, for the fourth scenario, we can work on MI signals with noise and without using the augmentation technique. The accuracy (Acc), sensitivity (Se),

Table 9.3: The Confusion matrix of ECG segments, **without noise** and with the 10-fold increase for **VGG-MI1**.

VGG-MI1	True/ Predicted	Normal	MI	Acc (%)	Se (%)	Pre (%)	Spe (%)
Without augmentation	Normal	20573	519	97.57	97.53	91.36	96.95
	MI	1945	78419	97.57	96.95	99.34	97.53
With augmentation	Normal	209174	1746	99.02	99.17	96.24	98.76
	MI	8168	795472	99.02	98.76	99.78	99.17

precision (Pre) and specificity (Spe) measures can be calculated for all these scenarios, and the comparison can be made between the proposed using the proposed sub-models (VGG-MI1 and VGG-MI2) and with the standard models such as the VGG and AlexNet.

In Tables 9.3, 9.4, 9.5, and 9.6, we show the results from the models in the form of the confusion matrices with all the folds. From the results, one can note that while using VGG-MI1, 97.53% of ECG signal segments with noise are correctly classified as the normal class while 96.95% of the ECG signal segments with noise are classified as MI. Similarly, we can see that 99.17% of ECG signal segments without noise are correctly classified as belonging to the normal class while 98.76% of the ECG signal segments without noise are classified correctly, having MI.

From Table 9.4, using VGG-MI1, with noise in the data and without using any data augmentation, we can also note that from 93.59% ECG signal segments are correctly classified as belonging to the normal class while 92.53% are correctly classified as having MI. Similarly, we note that 94.83% of ECG segments are correctly classified as belonging to the normal class, while 95.46% of ECG signal segments are correctly classified as MI with augmentation in place.

From Table 9.5, using VGG-MI2, with noise in the data and without using any data augmentation, we can also note that from 98.53% ECG signal segments are correctly classified as belonging to the normal class

Table 9.4: The Confusion matrix of ECG segments, **with noise** and with the 10-fold increase for **VGG-MI1**.

VGG-MI1	True/ Predicted	Normal	MI	Acc (%)	Se (%)	Pre (%)	Spe (%)
Without augmentation	Normal	19744	1352	92.75	93.59	76.68	92.53
	MI	6003	74361	92.75	92.53	98.21	93.59
With augmentation	Normal	200031	10889	95.33	94.83	84.59	95.46
	MI	36421	767219	95.33	95.46	98.60	94.83

Table 9.5: The Confusion matrix of ECG segments, **without noise** and with the 10-fold increase for **VGG-MI2**.

VGG-MI2	True/ Predicted	Normal	MI	Acc (%)	Se (%)	Pre (%)	Spe (%)
Without augmentation	**Normal**	20783	309	98.07	98.53	92.68	97.95
	MI	1640	78724	98.07	97.95	99.60	98.53
With augmentation	**Normal**	209856	1064	99.22	99.49	96.84	99.15
	MI	6828	796812	99.22	99.15	99.86	99.49

while 97.95% are correctly classified as having MI. Similarly, we note that 99.49% of ECG segments are correctly classified as belonging to the normal class, while 99.15% of ECG signal segments are correctly classified as MI with augmentation in place.

It can be observed that for the sub-model, VGG-MI1, a total of 2.4% of the ECG signal segments are wrongly classified as MI, and a total of 2.3% MI ECG signal segments are wrongly classified as belonging to the normal class—without using data the augmentation technique after removing the noise. Furthermore, a total of 0.8% normal ECG signal segments are wrongly classified as MI, and a total of 1% MI ECG signal segments are wrongly classified as belonging to the normal class using the augmentation technique after removing the noise.

Similarly, we can also observe for VGG-MI2, a total of 1.4% of normal ECG signal segments are wrongly classified as MI, while a total of 2% MI ECG signal segments are wrongly classified as belonging to the normal class—without using augmentation technique after removing the noise. Furthermore, a total of 0.5% of the normal ECG signal segments are wrongly classified as MI, while a total of 0.8% MI ECG signal segments are wrongly classified as belonging to the normal— using the augmentation technique after removing the noise.

As for measuring the accuracy of the CNN models proposed here, from Tables 9.7 and 9.8, we can see that for VGG-MI1 the

Table 9.6: The confusion matrix of ECG segments, **with noise** and with 10-fold increase using **VGG-MI2**.

VGG-MI2	True/ Predicted	Normal	MI	Acc (%)	Se (%)	Pre (%)	Spe (%)
Without augmentation	**Normal**	20152	940	95.17	95.54	83.60	95.08
	MI	3951	76413	95.17	95.08	98.78	95.54
With augmentation	**Normal**	204170	6750	97.24	96.79	90.64	97.35
	MI	21250	782390	97.24	97.35	99.14	96.79

Table 9.7: The overall classification results for **VGG-MI1**.

VGG-MI1	TP	TN	FP	FN	Acc (%)	Se (%)	Pre (%)	Spe (%)
With Augmentation	795472	209174	1746	8168	99.02	98.76	99.78	99.17
Without Augmentation	78149	20573	519	1945	97.57	96.95	99.34	97.53

accuracy, sensitivity, and specificity are 99.02%, 98.76%, and 99.17%, respectively. This was achieved with ECG signal segments with no noise and also with data augmentation in place. Similarly, for the VGG-MI2, the accuracy, sensitivity, and specificity is 9.22%, 99.15%, and 99.49%, respectively. Hence, we can conclude that the performance of the VGG-MI2 is slightly better when compared to the VGG-MI1 sub-model. This is because the QG-MSVM classifier introduced in the VGG-MI2 model appears to be playing an important role to increase the accuracy of the classification. Moreover, we note that the performance of VGG-MI2 with data and without data, augmentation performs better than the VGG-MI1 sub-model. In Table 9.9, we show the comparative study of the proposed two sub-models we have introduced here against the standard VGG and the AlexNet.

Thus, in terms of comparison, the deep learning framework, with smaller sub-models derived from the VVG framework appears to be very efficient in detecting MI using small segments of EEG signals.

In part, there are many methods applied to analyse the ECG signal for MI detection. For example, Sun et al. [12] proposed an automated MI detection system based on multiple instances of learning. Accordingly, they report results with sensitivity and specificity of 92.30% and 88.10%, respectively, using kNN ensemble as their chosen classifier. Similarly, Sharma et al. [13] proposed a technique based on multiscale energy and Eigenspace for the detection of MI with accuracy, sensitivity, and specificity of 96%, 93%, and 99%, respectively, using an SVM classifier, and also accuracy, sensitivity, and specificity of 81%, 85%, and 87%, respectively, using a kNN classifier. However, it is noteworthy

Table 9.8: The overall classification results using **VGG-MI2**.

VGG-MI2	TP	TN	FP	FN	Acc (%)	Se (%)	Pre (%)	Spe (%)
With Augmentation	796812	209856	1064	6828	99.22	99.15	99.86	99.49
Without Augmentation	78724	20783	309	1640	98.07	97.95	99.60	98.53

Table 9.9: Summary of the evaluation results of the proposed models and the other two models using the first and the second scenarios. Note, the results for the proposed approaches are given in bold.

Model	Acc (%)	Se (%)	Pre (%)	Spe (%)
VGG-Net with Augmentation	97.35	97.44	99.20	97.02
VGG-Net without Augmentation	97.11	97.25	99.07	96.54
Alex-Net with Augmentation	98.69	98.63	99.67	98.77
Alex-Net without Augmentation	98.24	98.03	99.73	99.01
VGG-MI1 with Augmentation	**99.02**	**98.76**	**99.78**	**99.49**
VGG-MI1 without Augmentation	**97.57**	**96.95**	**99.34**	**97.53**
VGG-MI2 with Augmentation	**99.22**	**99.15**	**99.86**	**99.49**
VGG-MI2 without Augmentation	**98.07**	**97.95**	**99.60**	**98.53**

that most of such work has either utilised limited datasets to validate the results and also does not have computational superiority. As a result, the deep learning-based approach proposed here appears to be somewhat more attractive.

Conclusions

The proposed deep learning model, discussed in detail in this chapter, can be considered to be an efficient mechanism for automatic detection and analysis of cardiac disorders. The novel approach utilised here is to take advantage of the existing structure of the VGG model and modify it to obtain two sub-models, namely the VGG-MI1 and VGG-MI2.

To test the proposed model, we have utilised ECG signals from the PTB dataset, which were converted to greyscale images. The images were augmented to extend the dataset by 10 fold. Experiments using the model and based on the dataset mentioned show that the proposed CNN framework is accurate and efficient. In fact, the results show that we achieve an accuracy of 99.22%. Furthermore, we report a sensitivity of 99.15% and a specificity of 99.49% when using the VGG-MI2. It appears that even in the worst-case scenario, the results come up with an accuracy of 97.24%. Furthermore, we report a sensitivity of 97.13% and a specificity of 96.53%, on VGG-MI2. To our knowledge, this appears to be the present state of the art in the field, showing the superiority of the deep learning models in signal-based optimised image analysis for diagnostic purposes.

References

[1] Cardiovascular disease. World Heart Day, [Online]. Available: http://www.who.int/cardiovascular_diseases/world-heart-day/en/. [Accessed: 31-August-2020].

[2] Thygesen, K., J. S. Alpert, A. S. Jaffe, M. L. Simoons, B. R. Chaitman, and H. D. White. Third universal definition of myocardial infarction. Circulation 126(16): 2020–2035, 2012.

[3] Sadhukhan, D., S. Pal, and M. Mitra. Automated identification of myocardial infarction using harmonic phase distribution pattern of ECG data. IEEE Transactions on Instrumentation & Measurement 99: 1–11, 2018.

[4] Jayachandran, E. S., K. P. Joseph, and U. R. Acharya. Analysis of myocardial infarction using discrete wavelet transform. Journal of Medical Systems 34(6): 985–992, 2010.

[5] Arif, M., I. A. Malagore, and F. A. Afsar. Detection and localisation of myocardial infarction using k-nearest neighbor classifier. Journal of Medical Systems 36(1): 279–289, 2012.

[6] Al-Kindi, S. G., F. Ali, A. Farghaly, M. Nathani, and R. Tafreshi. Towards real-time detection of myocardial infarction by digital analysis of electrocardiograms. 1st Middle East Conference on Biomedical Engineering, pp. 454–457, 2011.

[7] Wu, J. F., Y. L. Bao, S. C. Chan, H. C. Wu, L. Zhang, and X. G. Wei. Myocardial infarction detection and classification—A new multiscale deep feature learning approach. IEEE International Conference on Digital Signal Processing, pp. 309–313, 2017.

[8] Acharya, U. R., H. Fujita, O. S. Lih, Y. Hagiwara, J. H. Tan, and M. Adam. Application of deep convolutional neural network for automated detection of myocardial infarction using ECG signals. Information Sciences, pp. 415–416, 190–198, 2017.

[9] Liu, W., Q. Huang, S. Chang, H. Wang, and J. He. Multiple-feature-branch convolutional neural network for myocardial infarction diagnosis using electrocardiogram. Biomedical Signal Processing & Control, 45: 22–32, 2018.

[10] Srivastava, N., G. Hinton, A. Krizhevsky, I. Sutskever, and R. Salakhutdinov. Dropout: A simple way to prevent neural networks from overfitting. Journal of Machine Learning Research 15(1): 1929–1958, 2014.

[11] Bouvrie, J. Notes on Convolutional Neural Network, 2007.

[12] Sun, L., Y. Lu, K. Yang, and S. Li. ECG analysis using multiple instance learning for myocardial infarction detection. IEEE Transactions in Biomedical Engineering 59(12): 3348–3356, 2012.

[13] Sharma, M., S. Singh, A. Kumar, R. San Tan, and U. R. Acharya. Automated detection of shockable and non-shockable arrhythmia using novel wavelet-based ECG features. Computers in Biology and Medicine 115: 103446, 2019.

Further Reading

Abubakar, A., H. Ugail, and A. Bukar. Noninvasive Assessment and classification of human skin burns using images of Caucasian and African patients. Journal of Electronic Imaging 29(4): 041002, 2019.

Bukar, A., and H. Ugail. On automatic age estimation from facial profile view. IET Computer Vision 11(8): 650–655, 2017.

Bukar, A., H. Ugail, and D. Connah. Automatic age and gender classification using supervised appearance model. Journal of Electronic Imaging 25(6): 061605, 2016.

Bukar, A., and H. Ugail. A nonlinear appearance model for age progression. *In*: Hassanien, Aboul Ella, Oliva and Diego Alberto (eds.). Advances in Soft Computing and Machine Learning in Image Processing (Springer). Springer, 2017.

Bukar, A., and H. Ugail. Facial age synthesis using sparse partial least squares (the case of Ben Needham). Journal of Forensic Sciences 62(5): 1205–1212, 2017.

Elmahmudi, A., and H. Ugail. Deep face recognition using imperfect facial data. Future Generation Computer Systems 19: 213–225, 2019.

Ugail, H., M. Alzorgani, and S. Betmouni. A Deep Learning Approach to Tumour Identification in Fresh Frozen Tissues, 13th International Conference on Software, Knowledge, Information Management and Applications (SKIMA), 2019.

CHAPTER 10

Advances in Visual Computing through Deep Learning

In the past few years, there has been a flurry of positive developments in the area of deep learning. This wave of developments, of course, has caught up in the area of visual computing too. From basic image processing tasks to visual analytics, the field of visual computing has benefited immensely from the tide of deep learning. Among the deep learning techniques, the convolutional neural networks (CNNs) have been extensively studied and applied in solving problems related to visual computing. With the explosion of the data, availability of complex pre-trained CNN networks and models, and also with the ease by which solutions or models in one area can be adapted to solve the challenges in other areas, deep learning has become the de-facto state of the art and the number one choice for solving challenging tasks in visual computing.

Aside from the key examples provided in the previous chapters, there are many more examples one can draw to illustrate the flurry of developments in visual computing where deep learning has been the driving force. Here we outline some more examples to showcase the applications of deep learning in visual computing.

Visual Detection and Tracking

Detecting objects from natural scenes is an important problem in visual computing which is still not sufficiently well addressed. For example, for autonomous vehicles to be able to drive efficiently, fast, accurate tracking of objects of all manners is essential. The use of CNNs for use in object detection and tracking from say video frames can go back to the 1990s. The main challenge back then has been the lack of training data and the lack of trained models, in addition to the limited computational power. However, since 2012, the demonstration of tremendous success in using CNNs for the ImageNet challenge has brought deep learning to the forefront across academic research labs in the field.

In the case of object detection in video frames, earlier work built upon the identification of a suitable window slider across images so that the CNN classifier can return the relevant target information. However, in many cases, because of the complexity of the backgrounds, for example, the number of sliding windows can get fairly large and the computation complexities, as a result, increases.

A common and rather successful CNN based approach to object detection nowadays is the so-called region-based CNN (R-CNN). As the name suggests, the task of the R-CNN is to investigate a given video frame at a smaller scale with a region-based approach defined as separate classes. Thus, in a given CNN, there could be as many as 2000 classes defining separate regions which can then be classified using Support Vector Machines (SVMs) or Cosine Similarity (CS) measures. Other methods such as pooling based on spatial information and multi-region enhanced semantic segmentation techniques have also been recently introduced.

A more recent example of an object detection framework is the YOLO (you only look once) framework. This is a single pipeline based pre-trained model which can undertake detection and return a prediction class for multiple objects. Thus, YOLO appears to apply just one neural network to the full image. It does this by making use of just one forward propagation iteration enabling the object to be recognised by looking at it only once. Following that, it automatically divides the image into smaller regions. Based on specified bounding boxes, it can assign probabilities of a given object belonging to the region. The final result then can be computed using the weights from each of the bounding boxes to determine the final probability. Unlike some other CNNs (such as the VGG), YOLO can run in real-time to provide high accuracy in object detection problems.

Pose Recognition

Pose estimation deals with knowing the orientation of an individual in a scene. Pose estimation has many real-world applications. These include understanding the physiology and body movement of people, gait analysis as a biometric and, the design of realistic-looking robots.

As far as deep learning goes, DeepPose has been one of the very first applications of deep learning for estimating poses from video footage. DeepPose uses a CNN-based regression to analyse the joint coordinates.

The model itself is a cascade 7 layer CNN derived from the VGG-19 model whereby each body joint is represented, and in combination, the full pose can be estimated.

Similarly, Mask RCNN is the deep learning-based model that uses both semantic and instance segmentation for predicting the pose. The method starts by extracting several feature maps from an image using CNN. These feature maps are then used to obtain bounding boxes which are then used for segmentation and finally prediction. The Mask RCNN is essentially a top-down approach for making the prediction, where key joint points are first detected before estimating the final pose of the person.

As for object tracking, it relies heavily on representation mechanisms where the target object may be in a challenging environment. The change in viewpoint, illumination, and occultation during the motion of the object adds significant challenges. Many types of CNNs have been put forward for object tracking.

In detector and tracker-based models, for example, the features of a given object to be tracked are learned offline. The architectures of the CNNs themselves are enhanced through invariances. Here, both spatial and temporal information can be learned, and then tracking can be performed.

Other techniques, using CNN for visual tracking, include training multiple CNNs with specific kernels for each of them. This allows the individual CNNs to be biased towards specific parts of the object or specific parts of the scene to be tracked. Similarly, custom CNNs with target-specific saliency maps along with generative target appearance models have also been popular recently.

Action Recognition

The main challenge in analysing videos using deep learning techniques is that the videos themselves are both spatial and temporal entities. Deep learning methods such as the CNNs often are geared to represent two dimension signals in the form of images of certain pre-selected sizes. The introduction of the temporal dimension makes it harder for the CNNs to handle the problem. Nevertheless, deep learning methods have performed well and have shown their superiority for action recognition tasks based on video input.

One way researchers have attempted to address the temporal dimension is the introduction of 3D CNNs where the network is extended to the temporal dimension too. Other methods researchers have utilised include the method of fusion in which different feature maps are derived and fused to obtain the final classification. Similarly, traditional methods such as the optical flow combined with CNN features and the introduction of novel learning algorithms which combine CNNs with other types of machine learning modules have been utilised.

Image Processing and Classification

Efficient methods and techniques for image processing and classification is a key ingredient in the solutions to many visual computing problems. As such, many suitable methods and techniques for image processing have existed for decades. However, the breakthrough for image processing and classification came in 2012 when AlexNet achieved the best classification performance in the Large Scale Visual Recognition Challenge 2012 (ILSVRC2012) challenge. The challenge aimed to analyse and infer the content of photos of 10,000,000 human-labelled images depicting 10,000+ object categories in the ImageNet dataset. The essential task was to correctly identify the main objects that were present in the images.

Like many modern CNNs, the AlexNet works through the use of efficient data filters and the introduction of the category of hierarchies for the labelled classes in the network. This enables the network to be trained with fewer examples, yet achieving significantly higher accurate results. Since then, many CNNs architectures have been proposed in that a tree-like structure of classes where the relationship between different classes can be better understood and utilised for efficient and accurate classification.

Similarly, subcategory classification techniques, based on CNNs have become popular recently. In this type of architecture, they adopt a coarse-to-fine classification approach which adds efficiency to the classification techniques utilised. In such schemes, parts of the network itself can be trained on parts of the object which allows for better recognition. For example, previously, we have shown how it will be possible to train a face recognition algorithm with parts of the face, thus enabling the algorithm to be far more accurate and efficient. Research, in the past few years, in utilising parts of an object, has shown that accuracy, in general, improves by adding local level features to the CNN, during training.

With the basic binary image classification problem solved, via the application of deep learning techniques, the focus more recently has shifted towards creating pipelines that can undertake key image processing and classification within a single entity. For example, a CNN architecture can be created such that it can have functionality within the network specifically for the localised operations and for segmentation before a category classification could occur.

Similarly, researchers are attempting to apply visual attention focussed CNNs which can perform classification at a much more detailed level. For example, one part of the network can have more emphasis on the image in question at a patch-based level, while the other part of the network can look at the object at a finer level. These levels can then be combined to obtain high levels of accuracy in object classification.

Scene Labelling

The technique of labelling a scene is the identification of semantic classes within an object and to automatically analyse a given scene, i.e., for an outdoor scene, one might need to be able to automatically identify the sky, trees, plants, and any wildlife that may be present in the scene.

Deep learning has gained popularity for its capacity to carry robust scene labelling. The related CNNs can model the scenes so that multiple classes from a given scene can be identified and labelled correctly. Very often, the use of multi-scale CNNs is seen to play a crucial role in trying to address the complexities of scene labelling. Other methods include training CNNs for patch-based classification.

There has been much success in utilising pre-trained models such as the AlexNet for semantic-based object segmentation for classification applied in scene labelling. Often, in this type of attempt, the deep characteristics of the images of objects in a scene can be extracted using the AlexNet features, and classification can be performed using SVM. The use of pre-trained models offers a degree of advantage in that smaller datasets are needed for re-training, yet the capacity for generalisation does exist.

Visual Saliency Detection

The task of visual saliency detection is to simulate the human visual system for understanding complex scenes. The prime goal of saliency detection is to come up with a computer model that can successfully locate the motion-related salient objects in a video sequence. This is obviously a spatiotemporal problem. Multi-contextual information is a crucial apriori in visual saliency prediction, and CNNs are usually geared to handle such problems.

Common CNN techniques for visual saliency detection involve dividing the problem into two simpler problems whereby the local and global saliency contexts are handled separately. For example, a suitably crafted CNN model can assign a local saliency value at the pixel level, and object-level information can be applied in a global context by a deep fully-connected feed forward network.

Most of the deep learning-based video saliency detection methods so far have attempted to design and train separate networks to address this problem. For example, it is common to utilise the encoder-decoder network type models to address this problem. Here, the encoder network can be utilised to extract the 3D deep features from the input video while the decoder network can compute the saliency.

Discussions

Deep learning has been able to start addressing many of the challenging issues in visual computing. Over the years, we have witnessed deep learning taking centre stage as a prime de-facto technique for solving the fundamental challenges arising in the world of visual computing. Though we have seen this rather dramatic development in terms of the numerous applications and the impressive results, such developments in many ways have been incremental. For instance, let's take the example of automatic text identification, which often is referred to as optical character recognition (OCR), which has a long history.

Traditionally the OCR is performed in very constrained environments that are well lit and clean. However, the real world is much more sophisticated than that. For example, reading the number plate of a fast-moving car in a severe weather condition is a very real-world problem. Such problems require non-traditional computation techniques. Often in unconstrained environments, there exist large amounts of appearance variations that are not controllable. However, the rich feature representational capacity that modern deep learning methods such CNNs provide can be utilised to tackle this problem. In a typical text detection and recognition task, there are two basic stages. They are the detection and recognition stage. Both these stages can sufficiently be addressed today using CNNs.

Typically, for the task of text recognition, a CNN model would learn to crop text patches and non-text patches so that they can be successfully discriminated. The detected text can then be mapped to specific filters in the network architecture. Often a multi-scale image pyramid is utilised for this purpose. Other techniques involve obtaining via Maximally Stable Extremal Regions (MSER) which can be then filtered by the network. As for the recognition stage, CNN can extract rich visual features using suitably trained Softmax classifiers.

In 1917, the 18-year-old blind Londoner Mary Jameson demonstrated the Optophone—a device that can scan text (roughly one word per minute) to produce a tone that the blind can interpret. That was probably the beginning of OCR. From then, fast forward time by 75 years, in 1992,we had the Newton MessagePad, which was launched with the promise that it can perform handwriting recognition. Commercially speaking it was short-lived and was a failure. However, through sustained research into image processing, the technology kept improving at a fast pace. Today, we more or less have technology that can most probably recognise any number of letters in any language under wildly unconstrained conditions. Much of that technology hinges on deep learning as the critical subsidising device.

Further Reading

Girshick, R., J. Donahue, T. Darrell, and J. Malik. Rich feature hierarchies for accurate object detection and semantic segmentation. In Proceedings of the IEEE Conference on Computer Vision and Pattern Recognition (CVPR), pp. 580–587, 2014.

Long, J., E. Shelhamer, and T. Darrell. Fully convolutional networks for semantic segmentation. IEEE Transactions on Pattern Analysis 7 and Machine Intelligence (PAMI) 39(4): 640–651, 2017.

Olga Russakovsky, Jia Deng, Hao Su, Jonathan Krause, Sanjeev Satheesh, Sean Ma, Zhiheng Huang, Andrej Karpathy, Aditya Khosla, Michael Bernstein, Alexander C. Berg, and Li Fei-Fei. ImageNet large scale visual recognition challenge. International Journal of Computer Vision 115: 211–252, 2015.

Redmon, J., S. Divvala, R. Girshick, and A. Farhadi. You only look once: Unified, real-time object detection. In Proceedings of the IEEE Conference on Computer Vision and Pattern Recognition (CVPR), pp. 779–788, 2016.

Toshev, A., and C. Szegedy. Deeppose: Human pose estimation via deep neural networks. In: Proceedings of the IEEE Conference on Computer Vision and Pattern Recognition (CVPR), pp. 1653–1660, 2014.

Uijlings, J. R., K. E. van de Sande, T. Gevers, and A. W. Smeulders. Selective search for object recognition. International Journal of Conflict and Violence, (IJCV) 104(2): 154–171, 2013.

CHAPTER 11

Frontiers and Challenges in Deep Learning for Visual Computing

Today the field of deep learning is well developed, and its application in the field of visual computing is diverse. From face recognition to providing visual analytical guidance to autonomous vehicles, it is deeply embedded into the field of visual computing. It is true to say that deep learning has been the most successful supervised machine learning approach in recent times. The success of deep learning is fuelled by its ability to generalise complex problems and its semantic learning capacity. With such characteristics, and with the availability of sufficient training data and computational resources, deep learning seems to be able to take on tough challenges which traditional image processing and data analytics solutions could not undertake.

Mathematically, deep learning can be seen as a complex function mapping such that, $f : X \rightarrow Y$, where a relationship between a given piece of data and output features can be derived. This allows us to bring about a sufficient level of abstraction whereby high-level information or knowledge can be derived from the data. Thus, deep learning models possess a hierarchical pattern inferring capacity in which a small set of building blocks are connected to form a complex deep architecture capable of carrying out "learning". A key characteristic of deep learning is the existence of multiple hidden layers within the architecture. Similar to the neural network structure of the human brain, this parallel architectural network within deep learning enables to process information through many stages from which data is transformed and represented.

In this book, we have discussed many different types of deep learning architectures ranging from simple autoencoders to deep convolutional neural networks. We have also discussed how deep learning can be applied to solve practical problems ranging from human face recognition to the determination of the condition of the heart through ECG signal analysis.

Looking at the evolution of deep learning over the last few decades, the field of visual computing has started benefiting greatly from deep learning methods and techniques since 2006. Since then, researchers have been steadily demonstrating that deep learning can overcome a significant amount of challenges which the traditional image classification methods could not handle. A notable turning point for deep learning to be able to take the centre was the moment that ResNet won the ILSVRC 2015 challenge. Since then, many pre-trained Convolutional Neural Networks (CNNs) have been successfully executed. These include the VGGNet, GoogLeNet, and AlexNet.

Frontiers

There are many parallel directions at which the developments of deep learning can be taken further forward. In some cases, researchers are constantly looking at ways to make the existing networks deeper while in other cases, they are looking into ways to improve the various sub functionalities of the networks. These may include optimising the arrangement of various layers, construction of better activation, and Softmax functions.

In the case of convolution operations, one might want to look at improving the representability of the convolution layers. These may include methods for weight sharing to potentially reduce the number of weights. Other methods might include the introduction of hyperparameters to the convolutional layers to "dilate" it. This can be carried out by allowing zeros between various filters and may prove to be useful for solving repetitive tasks. Another way to look into re-arranging the convolutional layers is the introduction of the "network in a network" whereby traditional linear filters can be replaced by a network itself. This should help to reduce the number of weights within the network and at the same time, may prove to be efficient.

Within a deep learning architecture, the activation functions play a crucial role. Having an efficient activation function enables the network to perform optimally. In this sense, the ReLU is a vital activation function whose form can be subjected to experiments. For example, a leaky ReLU—which is derived by combining a max function and min function may allow enhancing the search space for better optimisation. Similarly, one could also look into non-linear ReLU units such as the exponential ReLUs. These are believed to bring enhanced learning and higher classification rates.

As far as looking at the loss function is concerned, it is essential to choose the right loss function for the specific task at hand. For example, rather than selecting a single parameter-based loss function, one could look into the triplet loss functions—which can consider three instances of the loss function. This could help in better classification for image and video processing tasks. Similarly, one could also look into combining the loss functions with the Softmax. Such transformations have the potential to obtain better visual classification results.

Deep learning is data-hungry—even for the simplest form of training—it demands large data. There is an imbalance between the number of parameters that are required to be correctly adjusted and the availability of sufficient data to enable this. One way to overcome this is by using data augmentation. The idea is to change the "shape" of the data without changing their meanings. Augmentation methods include linear transformation and various other types of photometric transformations.

Thus, a significant frontier in deep learning is the provisions available for further investigation into the network structures and look into more the individual components. Above, we have highlighted some areas where further research may lead to exciting findings with the potential to further enhance the power of deep learning.

As mentioned earlier, deep learning started becoming prominent in the early 90s, where the use of CNNs for handwriting recognition has shown to be becoming superior. Then the AlexNet architecture sparked interest among researchers in the field which was quickly followed by the VGG models, GoogLeNet, and the ResNet architectures. Much of these architectures and the problems that deep learning has successfully been able to tackle have been supervised machine learning problems. Such problems are classed as regular, requiring an input/output pipeline. There exist many problems which cannot be formulated to suit such a supervised learning framework. It would, for example, require the limited input data to be modified in specific ways so that the problem can be solved using more straightforward solutions.

Challenges

There is little doubt that deep learning has recently become very successful in solving many visual computing problems. Due to the agile nature of the network architectures, the availability of sufficient data and computational resources has helped with this feat. Even though deep learning has had success in the domain of visual computing, there are significant issues or challenges ahead that one must bear in mind.

As it stands, deep learning cannot be fully explained. Much of the deep learning techniques and models are run as black boxes. This means it is difficult to explain why and how deep learning would fail. Furthermore, deep learning is currently not inferential enough. Much of the deep architectures are prediction models which have limitations. More recently, researchers, as well as practitioners, have been interested in the idea of explainable AI (XAI) where appropriate explanations are required as to how and why the AI is making a particular decision and not any other. Due to the gaps in the theoretical understanding of deep learning, the task of applying XAI to it is currently a critical challenge.

Deep learning is data-hungry. Much of the success we have seen over the years for the use of deep learning in solving visual computing-related problems is the availability of sufficient training data. However, even for relatively small tasks, deep learning appears to be asking for large volumes of training data. Whilst data augmentation and other such tricks offer to relieve this burden to a certain degree, because of this fact, the practical applications of deep learning can be limited. This is simply because one could not provide sufficient data or case studies to train the current deep architectures which are in place. Hence this is another critical challenge one must bear in mind in working with deep learning algorithms.

Furthermore, hyperparameter optimisation is one challenging task when resorting to training a deep learning network from scratch. Usually, this is carried out by a specialist in the field, and even then, a trial and error approach is implemented for this task. Often it takes days for even a specialist to adjust the parameters to tune the model so that it can return optimal results. Thus, developing intuitive parameter adjustment techniques and methods is also one challenging task.

One basic assumption researchers take when implementing deep learning models is that the problem at hand is relatively well defined and contained. In other words, the problem is often transformed into a domain where it does not interfere with otherworldly issues. However, no practical problem can receive such privileged exclusivity. In a sense, when implementing deep learning, we assume that the world we live in is largely stable and inferable from a given dataset. However, that is far from the reality of the case.

Index

About the Author

Professor Hassan Ugail is the director of the Centre for Visual Computing at the University of Bradford in the UK. He is a mathematician and a renowned computer scientist in the area of visual computing and artificial intelligence.

https://en.wikipedia.org/wiki/Hassan_Ugail

tweet @ugail

https://www.facebook.com/ugail

Printed in the United States
by Baker & Taylor Publisher Services

Printed in the United States
by Baker & Taylor Publisher Services